Nature and Origin of Cretaceous Carbon-rich Facies

17

Nature and Origin of Cretaceous Carbon-rich Facies

Edited by

S. O. Schlanger

Department of Geological Sciences, Northwestern University, Evanston, Illinois, USA

M. B. Cita

Institute of Paleontology, University of Milan, Milan, Italy

 1982

ACADEMIC PRESS

A Subsidiary of Harcourt Brace Jovanovich, Publishers

LONDON   NEW YORK
PARIS   SAN DIEGO   SAN FRANCISCO
SÃO PAULO   SYDNEY   TOKYO   TORONTO

ACADEMIC PRESS INC. (LONDON) LTD.
24/28 Oval Road
London NW1

United States Edition published by
ACADEMIC PRESS INC.
111 Fifth Avenue
New York, New York 10003

British Library Cataloguing in Publication Data
Nature and origin of Cretaceous carbon-rich facies
  1. Geology, Stratigraphic—Carbonferous
  I. Schlanger, S. O.   II. Cita, M. B.
  551. 7′5   QE671
  ISBN 0–12–624950–4

Typeset in Great Britain by Latimer Trend & Company Ltd, Plymouth
Printed in Great Britain by Taylor & Francis (Printers) Ltd, Basingstoke.

List of Contributors

M. A. Arthur   Department of Geology, University of South Carolina, Columbia, South Carolina, 29208, USA

W. B. N. Berry   Department of Paleontology, University of California, Berkeley, California, 94720, USA

P. de Boer   Institute of Earth Sciences, State University of Utrecht, Budapestlaan 4, Utrecht, Netherlands

H. Chalmley   Sedimentologie et Geochimie, Université de Lille, Lille, France

M. B. Cita   Institute of Paleontology, University of Milan, Piazzale Gorini 15, Milan, Italy

G. E. Claypool   U.S. Geological Survey, Denver, Colorado, 80225, USA

W. E. Dean   U.S. Geological Survey, Denver, Colorado, 80225, USA

J. V. Gardner   U.S. Geological Survey, Menlo Park, California, 94025, USA

D. Grignani   AGIP-SGEL, San Donato, Milan, Italy

D. Habib   Department of Earth and Environmental Sciences, Queens College, Flushing, New York, 11367, USA

J. McKenzie   Geological Institute, ETH, Sonneggstrasse 5, Zurich, Switzerland

C. Robert   Géologie Marine, Université d'Aix Marseille, Marseille, France

S. O. Schlanger   Department of Geological Sciences, Northwestern University, Evanston, Illinois, 60201, USA

I. Premoli Silva   Institute of Paleontology, University of Milan, Piazzale Gorini 15, Milan, Italy

B. R. T. Simoneit   School of Oceanography, Oregon State University, Corvallis, Oregon, 97331, USA

D. H. Steurmer   Lawrence Livermore National Laboratory, University of California, Livermore, California, 94550, USA

J. Thiede   Department of Geology, University of Oslo, P.O. Box 1047, Blindern, Oslo, Norway

P. Wilde   Lawrence Berkeley Laboratory, University of California, Berkeley, California, 94720, USA

Preface

Among the many problems engaging the attention of IGCP project 58 "Mid-Cretaceous Events", one of the foremost has been the question of relationships between paleoclimate, marine sedimentation and the chemical environment of the sea. The subject of anoxic sediments, that is, sediments generally considered to have formed in waters deficient in, or lacking, oxygen, is one that has interested geologists for many years. The collection of papers presented in this volume, which is appropriately entitled "Nature and Origin of Cretaceous Carbon-rich Facies" for reasons which will become apparent from the contents of the contributions, takes us a giant stride forward towards understanding the genesis of these sediments. Furthermore, it is eloquently demonstrated that there is a clear association between oceanic anoxic events and the concentration of hydrocarbons in the Cretaceous System, a fact of considerable significance for petroleum exploration.

The papers presented in this volume constitute a synthesis and analysis of a symposium organized by Professors Cita and Schlanger, for the working group on "Anoxic Events" of Mid-Cretaceous Events, and held in conjunction with the 26th International Geological Congress in Paris (1980). It represents an updating of the results presented at the symposium with due regard to the scientific discussions.

On behalf of the project working group of Mid-Cretaceous Events, I thank Professors Cita and Schlanger for the time and effort they have put into preparing the material for publication and the contributors for their willingness to make the work of the project such a success.

Richard Reyment
Leader of IGCP Project 58 "Mid-Cretaceous Events"
August 1982

Contents

1. Introduction to the Symposium "On the Nature and Origin of Cretaceous Organic Carbon-Rich Facies"

S. O. Schlanger

Department of Geological Sciences, Northwestern University, Illinois

M. B. Cita

Institute of Paleontology, University of Milan

At the Third International Congress of the "Mid-Cretaceous Events" Project of the IGCP held at Nice, France in October 1976, a round table discussion focused on the problem of anoxic marine sedimentation. This focus had been prompted by the increasing availability of evidence, gathered largely from Deep Sea Drilling Project cores, indicating that organic carbon-rich sediments were a global component of mid-Cretaceous strata in the major ocean basins, on oceanic plateaus, in shelf settings, and in epicontinental seas (Lancelot *et al.*, 1972; Berger & von Rad, 1972; Jackson & Schlanger, 1976). Further, a correlation between the "Oceanic Anoxic Events" of Schlanger and Jenkyns (1976) and the concentration of hydrocarbons in Cretaceous strata, as discussed below, was becoming evident.

A number of geologists, paleontologists, sedimentologists and paleoceanographers had, at the time of the Congress, already prepared papers reviewing the causes of, and indeed our state of ignorance concerning, anoxic sediments in the world ocean (e.g. Schlanger & Jenkyns, 1976; Fischer & Arthur, 1977; Ryan & Cita, 1977; Thiede & van Andel, 1977). In 1976 the issue was not a new one. Hallam (1967) had reviewed the subject of the depth significance of bituminous, laminated shales in the light of oceanic organic productivity and the position of the oxygen minimum zone. Frush and Eicher (1975) and Simpson (1975) had invoked water column anoxia to explain faunal and sedimentological features of Cretaceous strata in the US and Canadian Western Interior Basins. However, the flood of new evidence arising by the middle 1970s, both from Deep Sea Drilling Project cores and from outcrop sections being increasingly visited by geologists affiliated with the Deep Sea Drilling Project, and the availability of dramatic new tools such as organic geochemical indicators of oxic v anoxic environments (see Didyk *et al.*, 1978) prompted the group at Nice to form Working Group 8 of the IGCP "Mid-Cretaceous Events" Project. The aim of this new working group was the fostering of research on organic carbon-rich strata. To this end a symposium, "Nature and Origin of Cretaceous Organic Carbon-Rich Facies", was planned for the XXVI International Geological Congress to be held in Paris in 1980. At the Paris Congress the symposium was held with S. O. Schlanger and M. B. Cita as co-conveners under the auspices of the "Mid-Cretaceous Events" Project. Fourteen talks were held dealing with paleontological, organic, isotope and inorganic geochemical, sedimentological and paleoenvironmental aspects of such strata, ten of these are presented in this volume. In order to broaden the geological time perspective of the symposium, papers on Paleozoic, Cenozoic and modern deposits were invited.

The first three papers deal with descriptions and interpretations of Cretaceous strata from the Tethys, the continental margin of north-west Africa, and from the submarine rises and plateaus of the Pacific Basin. Arthur and Premoli Silva point out that the Cretaceous "Oceanic Anoxic Events" of Schlanger and Jenkyns (1976) are expressed in the pelagic sequences of the Tethyan seaway. In the Umbrian region and in the southern Alps these sequences are characterized by a rhythmicity of the order of 20 000 to 50 000 years. The lithologic, paleontologic, and chemical relations suggested to them that periodic upwelling was the originating factor in the formation of the prominent rhythmic bedding. The water structure was probably stable and oxygen contents relatively low. Their data do not allow them to distinguish between an expanded oxygen-minimum model and an entirely anoxic or dysaerobic deep-basin model. They also interpret the short-lived Cenomanian–Turonian anoxic event represented by the Bonnarelli Bed of the Appenines as the result of an intense major upwelling event. Dean and Gardner who also studied cyclically bedded units in Deep Sea Drilling Project cores taken off the coast of north-west Africa argue that the high input of turbidite derived organic matter transported from the shallower oceanic regions caused bottom water anoxia; the preservation of organic matter in these sediments therefore did not require original bottom water anoxia. Their argument are impressive for the Aptian–Albian cyclically bedded strata where black beds are enclosed by oxic sediments. However, the short-lived Cenomanian–Turonian Bonnarelli Bed type of deposit which is rich in marine-derived organic matter may not be amenable to this argument of turbidite control. Thiede, Dean and Claypool attacked the problem of the deposition of organic-rich strata on the relatively isolated, high-standing oceanic rises and plateaus in the Pacific Basin such as the Mid-Pacific Mountains, the Hess Rise, the Shatsky Rise, and the Manihiki Plateau. These authors do not deem it necessary to invoke an extensive mid-water oxygen deficiency or general euxinic conditions on a global scale to account for the Pacific occurrences. They argue that each of the Pacific occurrences of mid-Cretaceous "anoxic" strata was preserved due to the coincidence of several tectonic and oceanographic factors. Further, they present two scenarios—"the organo-volcanic spike" and the "equatorial crossing" models to account for all of the organic carbon-rich layers in the Pacific Cretaceous.

The next group of four papers explores the mineralogical, paleontological, and geochemical aspects of this facies. Chamley and Robert discuss clay mineralogy of black shales from the Atlantic continental margins. The detrital nature of the clays permits reconstruction of environmental conditions of source areas. The abundance of smectite suggests to the authors badly drained coastal areas, hence tectonic stability and low-relief morphology. A progressive development of longitudinal currents in the Late Cretaceous is supported by the diversification of terrigenous supply.

Habib postulates a "sedimentary supply origin" for the Cretaceous black shales from the Atlantic Basin. His palynological studies lead him to the conclusion that the rate of supply of organic matter and the rate of supply of other sediments, if high enough, could lead to the formation of the black shale facies regardless of the degree of oxicity of the bottom waters. Early Cretaceous deltaic systems would have supplied the large amounts of terrigenous plant debris.

De Boer presents an isotopic study carried out on Middle Cretaceous pelagic sediments from the central Apennines. Stable oxygen isotope data point to regular changes in temperature. These changes are accounted to a regularly north- and southward shifting of the caloric equator and the tropical upwelling zone.

Simoneit and Steurmer bring to our attention the fact that over the past decade the powerful tool of organic molecular characterization has been applied to the problem of determining the source of the organic matter in marine sediments, i.e. terrigenous plants v. marine planktic algae and animals and the degree of oxicity of the paleoenvironment. They review the literature and provide data and interpretations on the kinds of molecular markers, particularly of the lipid group, that are useful in the interpretations of the sources and depositional environments of organic carbon-rich sediments and their hydrocarbon potential.

The final group of three papers reports on the results of studies of modern, Cenozoic and Paleozoic strata. Cita and Grignani briefly review Late Neogene Mediterranean sapropels within a precise stratigraphic framework. Kerogen composition indicates a marine origin for all these Mediterranean sapropels (from DSDP cores and from piston-cores) which can be explained by the euxinic model. Different mechanisms seem to be able to produce basin-wide stagnation, as revealed by the different response of planktonic and benthic populations.

McKenzie presents an isotopic study carried out on a core raised from Lake Greifen, a small Swiss lake which has become progressively eutrophic in the last 100 years. The historical process of eutrophication is traced using changes in the carbon-13 content of the lacustrine chalks. The lake model is then used to explain the carbon-13 enrichment in pelagic carbonates during the Early and Middle Cretaceous.

Wilde and Berry, having studied the occurrences of Paleozoic black shales in cratonic settings, develop a physical oceanographic model based on considerations of water temperature, salinity, evaporation–precipitation ratios and oxygenation opportunities of oceanic waters. They sketch scenarios, using climatic controls, in which the world ocean, in Mesozoic time, could have reverted to a state resembling the Early Paleozoic in terms of widespread anoxia.

For many years, petroleum geologists have considered that dark, basinal shales, in a close facies association with potential reservoir rocks, were the source beds of many major hydrocarbon accumulations. However, it has only been in the past few years, with advances in organic geochemistry and thermal maturation studies, that the importance of the environment of deposition and the primary source of organic carbon in these shales has been recognized as a controlling factor in the generation of oil and gas (e.g. Tissot et al., 1980; Dow, 1978). Further, with the accumulation of oil and gas statistics, it has been increasingly apparent that Cretaceous strata are unusually prolific oil and gas producers. Also, with our recognition of Cretaceous periods of widespread, indeed global, oceanic oxygen deficiencies it has become clear that there is a causal link between oceanic anoxia and source bed formation (Arthur & Schlanger, 1979; Demaison & Moore, 1980). Studies of petroleum production and reservoir statistics have demonstrated the prolific nature of Cretaceous strata (Halbouty et al., 1970; Klemme, 1975; Moody, 1975). In a review of the distribution of oil in time, Tiratsoo (1973) pointed out that 63% of the world's giant oil fields have reservoirs of Mesozoic age, whereas 24% have Tertiary reservoirs, 13% have Paleozoic reservoirs, and 75% of all produced oil has come from giant fields. Exclusion of the Middle East fields changes these figures to 40% Tertiary, 39% Mesozoic, and 21% Paleozoic. However, exclusion of the Middle East fields obscures the fact that of the 250 commercial reservoirs in the region described up to 1967 (Tiratsoo, 1973), more than 200 are Middle Jurassic to Late Cretaceous in age. Further, Early to Middle Cretaceous strata are thought to be the source beds for many of the Middle East fields.

Irving et al. (1974) pointed out that over 70% of all known pooled oil is 60–140 m.y. old (Tithonian through Danian). They also pointed out that within this interval, 60% of the world's known oil is between about 110 and 80 m.y. old (Albian through Coniacian). This latter interval represents only 5% of Phanerozoic time, but oil apparently accumulated during this time interval at 12 times the Phanerozoic average (Irving et al., 1974). These writers postulated that the occurrence of much mid-Cretaceous oil was due to a combination of climatic, oceanographic, and tectonic factors. Even present figures may underestimate the relative abundance of Cretaceous oil; Young *et al.* (1977) demonstrated that several reservoirs of Miocene age in the northern Persian Gulf and the Lake Maracaibo basin contain oil of Cretaceous age.

Two studies by North (1979) and Tissot (1979) are particularly instructive. These authors compiled data on the age distribution of oil source beds (Figure 1). The overwhelming preponderance of Jurassic and Cretaceous strata as source beds makes it obvious that the Jurassic and Cretaceous seas were particularly prone to the sequestering of organic carbon in marine sediments. The picture of the time distribution of natural gas is not quite so clear. North (1977) in a review of the age incidence of natural gas pointed out that "The greatest single concentration of nonassociated natural gas known in the world is in the northwestern Siberian Basin and virtually all of it is in middle Cretaceous reservoirs" (p. 704). However, he considered the West Siberian Basin to be ". . . an exceptional case." A recent study by Masters (1979), however, argues for the recognition of ". . . an enormous tight-sand gas trap in western Canada" (p. 152). He also points out (p. 155) that "it must be significant that three of the largest sandstone gas fields in western NA are in low porosity Cretaceous ss. . . ." and further (p. 176) that "It is interesting although not critically important that the Basin Dakota, Wattenberg Dakota J, and Deep Basin Cadotte sandstones are all basal Late Cretaceous in age." Masters (1979, p. 176) also emphasized that "These named fields are the larger of hundreds of oil and gas fields in the Dakota–Muddy–Viking–Cadotte continental-shoreline–shallow-water com-

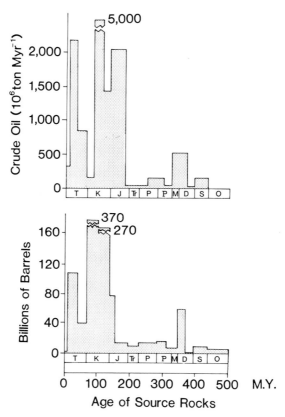

Figure 1. Age of petroleum source beds plotted against amount of oil per stratigraphic interval (upper graph from Tissot, 1979, lower graph from North, 1979).

plex of sandstones which extends for 1500 miles (2415 km) from the San Juan Basin in New Mexico to northwestern Alberta." The stratigraphy and paleogeography of the Western Siberian Basin and the US and Canadian Western Interior Basins are strikingly similar. Both of these basins, when viewed on a north polar projection of a plate reconstruction of 100 m.y. ago are seen to have been shallow epicontinental extensions of the north polar sea. It is not unreasonable to argue that in addition to stratal and age similarities both of these basins were similar in a paleoenvironmental context (i.e. they both contained, during the Cretaceous, anoxic water layers that tended to preserve organic carbon (Byers & Larson, 1979; Kauffman, 1977)). With the recognition that Cretaceous "Oceanic Anoxic Events" took place during relatively discrete time periods arguments have now been put forward that the coincidence between these events and the abundance of oil and gas in Cretaceous strata reveals a causal connection (Arthur & Schlanger, 1979; Scholle & Arthur, 1980; Demaison & Moore, 1980; Tissot et al., 1980). Not only true oceanic basins were affected by anoxia during the Cretaceous but the major epicontinental seas also were. The inability of much of the world ocean to oxidize the input of organic carbon during "Oceanic Anoxic Events" led to the eventual accumulation of much of the world's oil and gas reserves. The need for a continued study of mid-Cretaceous strata and paleoenvironments is obvious.

Acknowledgments

The co-conveners and authors wish to thank the following for their constructive reviews of the prepared papers: G. Eaton, BP, UK; D. Locke, Queens College, UK; W. Reed, Indiana University, ETH, Switzerland; J. S. Leventhal, USGS, USA; C. Weaver, Georgia Institute of Technology, USA; H. C. Jenkyns, Oxford University, UK. Miss E. A. Zbinden of the University of Hawaii helped prepare some manuscripts for publication. Special thanks are due to Professor R. A. Reyment of Uppsala University, Sweden, Project Leader of "Mid-Cretaceous Events" for his encouragement and support. C. Cheverton prepared the Index.

References

Arthur, M. A. & Schlanger, S. O. 1979. Cretaceous "oceanic anoxic events" as causal factors in development of reef-reservoired giant oil fields. *American Association of Petroleum Geologists Bulletin* **63**, 870–885.

Berger, W. & von Rad, U. 1972. Cretaceous and Cenozoic sediments from the Atlantic Ocean. *Initial Reports of the Deep Sea Drilling Project* **14**, 784–954.

Byers, C. W. & Larson, D. W. 1979. Paleoenvironments of Mowry shale (Lower Cretaceous), western and central Wyoming. *American Association of Petroleum Geologists Bulletin* **63**, 354–361.

Demaison, G. J. & Moore, G. T. 1980. Anoxic environments and oil source bed genesis. *American Association of Petroleum Geologists Bulletin* **64**, 1179–1209.

Didyck, B. M., Simoneit, B. R. T., Brassell, S. C. & Eglinton, G. 1978. Organic geochemical indicators of palaeoenvironmental conditions of sedimentation. *Nature* **272**, 216–222.

Dow, W. 1978. Petroleum source beds on continental slopes and rises. 1978. *American Association of Petroleum Geologists Bulletin* **61**, 423–442.

Fischer, A. G. & Arthur, M. A. 1977. Secular variations in the pelagic realm. *Society of Economic Paleontologists and Mineralogists, Special Publication* **25**, 19–50.

Halbouty, M. T. *et al.* 1970. World's giant oil and gas fields, geologic factors affecting their formations and basin classification, in Geology of giant petroleum fields. *American Association of Petroleum Geologists Memoir* **14**, 502–555.

Hallam, A. 1967. The depth significance of shales with bituminous laminae. *Marine Geology* **5**, 481–494.

Irving, E., North, F. K. & Couillard, R. 1974. Oil, climate and tectonics. *Canadian Journal Earth Sciences* **11**, 1–15.

Jackson, E. D. & Schlanger, S. O. 1976. Regional synthesis, Line Islands chain, and Manihiki Plateau, Central Pacific Ocean, DSDP Leg 33. *Initial Reports of the Deep Sea Drilling Project* **33**, 915–927.

Jenkyns, H. C. 1980. Cretaceous anoxic events from continents to oceans. *Journal Geological Society of London* **137**, 171–188.

Kauffman, E. G. 1977. Geological and biological overview: western interior Cretaceous basin. *The Mountain Geologist* **14**, 75–99.

Klemme, H. D. 1975. Giant oil fields related to their geologic setting; a possible guide to exploration. *Bulletin Canadian Petroleum Geology* **23**, 30–66.

Masters, J. A. 1979. Deep basin gas trap, western Canada. *American Association of Petroleum Geologists Bulletin* **63**, 152–181.

Lancelot, Y., Hathaway, J. C. & Hollister, C. D. 1972. Lithology of sediments from the western north Atlantic, Leg 11, Deep Sea Drilling Project. *Initial Reports of the Deep Sea Drilling Project* **11**, 901–949.

Moody, J. D. 1975. Distribution and geological characteristics of giant oil fields. *Petroleum and Global Tectonics*. Princeton, New Jersey: Princeton University Press, pp. 307–320.

North, F. K. 1977. Age incidence of natural gas. *Bulletin Canadian Petroleum Geology* **25**, 704–706.

North, F. K. 1979. Episodes of source-sediment deposition. *Journal of Petroleum Geology* **2**, 199–218.

Ryan, W. B. F. & Cita, M. B. 1977. Ignorance concerning episodes of ocean-wide stagnation. *Marine Geology* **23**, 197–215.

Schlanger, S. O. & Jenkyns, H. C. 1976. Cretaceous oceanic anoxic events: causes and consequences. *Geologie en Mijnbouw* **55**, 179–184.

Scholle, P. A. & Arthur, M. A. 1980. Carbon isotope fluctuations in Cretaceous pelagic limestones: potential stratigraphic and petroleum exploration tool. *American Association of Petroleum Geologists Bulletin* **64**, 67–87.

Simpson, F. 1975. Marine lithofacies and biofacies of the Colorado Group (middle Albian to Santonian). *Saskatchewan. The Geological Association of Canada Special Paper* **13**, 553–587.

Tiratsoo, E. N. 1973. *Oil Fields of the World*. Beaconsfield, England: Scientific Press.

Tissot, B. 1979. Effects on prolific petroleum source rocks and major coal deposits caused by sea level changes. *Nature* **277**, 463–465.

Tissot, B., Demaison, G., Masson, P., Delteil, J. R. & Combaz, A. 1980. Paleoenvironment and petroleum potential of middle Cretaceous black shales in Atlantic basins. *American Association of Petroleum Geologists Bulletin* **14**, 2051–2063.

Young, A., Monaghan, P. H. & Schweisberger, R. T. 1977. Calculations of ages of hydrocarbons in oils—physical chemistry applied to petroleum geochemistry I. *American Association of Petroleum Geologists Bulletin* **61**, 573–600.

2. Development of Widespread Organic Carbon-Rich Strata in the Mediterranean Tethys

M. A. Arthur

Department of Geology, University of South Carolina

I. Premoli Silva

Institute of Paleontology, University of Milan

Carbonaceous sediments of Early to Middle Cretaceous age are common to sedimentary sequences of the Tethys region now exposed in Alpine Mountain belts. Barremian through Albian pelagic sediments, relatively rich in organic carbon (Corg) were deposited on shallow pelagic plateaus, slopes and on oceanic crust along the southern margin of Tethys. Rhythmically interbedded, bioturbated pelagic carbonates and dark, laminated carbonaceous shales suggest conditions of periodic anoxia in deep water masses and fluctuating productivity or dissolution of carbonate within a period of 20 000 to 50 000 years. Corg contents of the organic carbon-rich strata average about 1.0%, but may be locally as high as 15.6%. A relatively small proportion of the preserved organic matter is terrestrially derived. The pelagic and hemipelagic sediments of the Early Cretaceous Tethys are similar to those in the North Atlantic.

A rise in carbonate dissolution levels and/or a fluctuating terrigenous clastic influx led to the deposition of shales and marlstones of Late Aptian–Middle Albian age following deposition of Tithonian–Aptian light-gray, cherty pelagic limestones. Late Albian and Cenomanian pelagic carbonates reflect an apparent drop of the calcite compensation depth or reduction of clastic influx, and an amelioration of the previous low oxygen conditions; thin carbonaceous layers occur. The general poorly oxygenated conditions in the Tethys seaway culminated in the Late Cenomanian–Early Turonian with deposition of up to 1 m of phosphatic, radiolarian muds rich in Corg (up to 23%) in upper slope and some shelf settings. The Corg appears to be entirely of marine origin and is hypothesized to have resulted from a sudden increase in organic productivity caused by overturn of nutrient-rich intermediate or deep water. Post-Turonian strata reflect better oxygenated conditions at the sea floor except in some deep basins along the Southern Alps. Sediments of equivalent age along the northern margin of Tethys also reflect anoxic conditions but are characterized by more detrital facies.

1. Introduction

Lower to Middle Cretaceous pelagic and hemipelagic sedimentary rocks rich in organic carbon have been the focus of much attention in the last few years because of their potential as hydrocarbon source rocks in deeper basins and along continental margins (e.g. Arthur & Schlanger, 1979; Demaison & Moore, 1980) and because they are globally widespread and occur in a variety of paleoenvironments and paleodepths. Many of the occurrences of these rocks and models for their deposition have recently been reviewed by Schlanger & Jenkyns (1976), Ryan & Cita (1977), Thiede & van Andel (1977), Fischer & Arthur (1977), Dean *et al.* (1978), Arthur & Natland (1979), Berger (1979), Thierstein (1979), Arthur & Schlanger (1979), Jenkyns (1980), Weissert (1981) and others, including papers in this volume. The

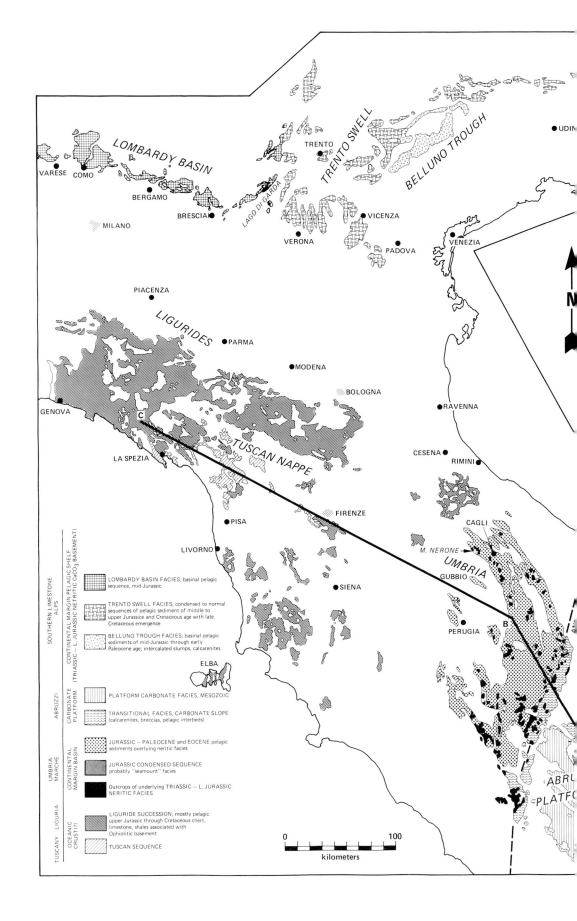

Figure 1. Geologic map of Northern Italy showing outcrops of late Mesozoic-Paleogene pelagic sediments and platform facies of Abruzzi. Cross-section ABC shown in Figure 2. Data from Ogniben (1975).

REA SHOWN ON MAP

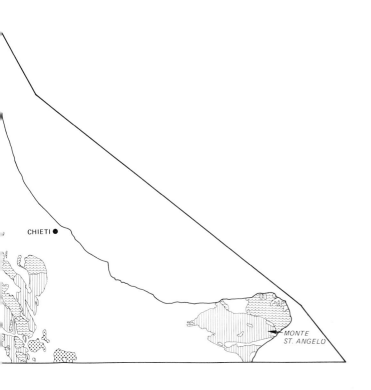

CHIETI ●

MONTE
ST. ANGELO

reader is referred to these papers for an overview of the problem of "oceanic anoxic events" and the widespread occurrence of organic carbon-rich strata.

The purpose of this paper is to present new data on the sedimentology, paleontology, chemistry, and organic content of so-called "black shale" units in Cretaceous Mediterranean Tethyan pelagic sequences. As will be shown, these are equivalent to the globally widespread "anoxic events", or periods of low-oxygen conditions, in much of the world ocean. Most of the discussion will center on the pelagic sedimentary rocks of Umbria in Italy, particularly the Gubbio section, and of the Southern Limestone Alps (Figure 1); the data are used to document the timing and duration of organic carbon-rich intervals and to constrain possible models for their origin. Trends in sedimentation in most Tethyan pelagic sequences are similar to those in the North Atlantic, pointed out by Bernoulli & Jenkyns (1974) and Jenkyns (1980).

2. Geologic setting

The pelagic sequences from Italy discussed here were deposited in a variety of environments, ranging from shallow continental plateaus to those in deeper basins on oceanic crust. The Cretaceous pelagic sequences in the Southern Alps were deposited mainly on continental crust and cap Early Mesozoic, mainly Triassic–Early Jurassic, carbonate platform sediments in a province of complex basin and swell topography (e.g. Hsü, 1976; Weissert, 1979; Jenkyns, 1980; Winterer & Bosellini, 1981). The location of some of these features is shown in Figures 1 and 2. The southern Alpine sections shown in Figure 3 represent pelagic sediments deposited both on submarine highs (e.g. S. Giacamo, Spiazzi, Pra da Stua) and in adjacent troughs (e.g. Breggia, Torre dé Busi). The basement of the Umbrian Apennines is also continental, and the Jurassic through Paleogene pelagic succession again overlies Triassic to Lower Jurassic platform carbonates. In Early Jurassic time, these regions were affected by extensional faulting and general subsidence, probably related to the opening of a portion of the Tethys seaway and formation of oceanic crust between the northern margin of the Apulian block, of which northern Italy was the western or north-western margin, and northern Europe (Smith, 1971; Dewey et al., 1973; Channell & Horvath, 1976; Winterer & Bosellini, 1981). The outer part of the margin subsided while some areas remained as blocks near sea-level where shallow-water carbonate sediments continued to accumulate. An example of this is the Abruzzi platform (Figure 1; d'Argenio et al., 1975; Colacicchi & Praturlon, 1965; Praturlon & Sirna, 1976) which lies to the south-east of Umbria. The hypothetical cross-section from the Abruzzi platform through the Ligurides to the north-west, shown in Figure 2, is not to scale, but is intended to show the distribution, sedimentary environments and approximate paleodepths of the different facies and provinces of northern Italy discussed here. The section is based mainly on inference from the composition and lateral changes in sediment of equivalent age described in the literature (see especially Bortolotti et al., 1970; Ogniben et al., 1975) or visited in the field during our work in Italy. Praturlon & Sirna (1976) and Bernoulli et al. (1979) have discussed facies relations on the Abruzzi platform and in the redeposited forereef sediments (Sabina facies) which interfinger with the pelagic Umbria and Marche facies, and their review is the basis for that portion of the interpretation in Figure 2.

Umbria is part of the Apulian block (Channell & Horvath, 1976), or Adriatic promontory, which moved with Africa relative to northern Europe. Paleolatitudes derived from paleomagnetic studies in Umbria (Vandenberg et al., 1978) suggest that the position of the Apulian block was between 20° and 30° North during most of the

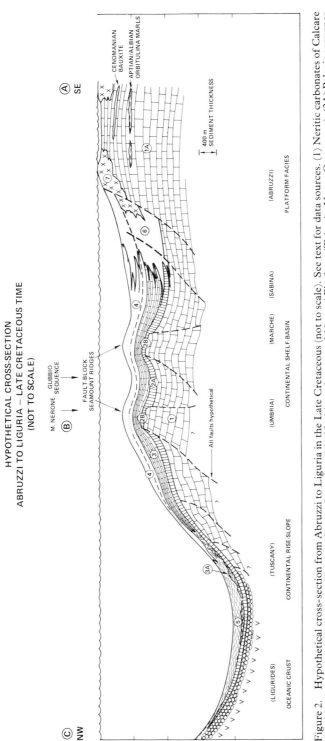

Figure 2. Hypothetical cross-section from Abruzzi to Liguria in the Late Cretaceous (not to scale). See text for data sources. (1) Neritic carbonates of Calcare Massicio (subsided platform) (Triassic–Lower Jurassic). (1A) Neritic carbonates of Abruzzi Platform (Triassic–Upper Cretaceous). (2A) Pelagic sequence (Lower–Upper Jurassic). (2B) Reduced pelagic seamount sequence (Lower–Upper Jurassic). (3) Maiolica Limestone and equivalent facies (Tithonian–Neocomian). (3A) Slumped Maiolica and breccias, slope and base of slope. (4) Scisti a Fucoidi and Scaglia in Umbria, Scisti a Policromi in Tuscany (Aptian–Upper Cretaceous). (5) Palombini Shale, Lavagna Shales, Scisti a Policromi and interbedded sand-calcarenite facies overlying ophiolite (Neocomian–Upper Cretaceous). (6) Sabina facies: redeposited shallow water carbonate interbedded with pelagic sediment (Jurassic–Cretaceous). (7) Rudist reefs (Middle–Upper Cretaceous).

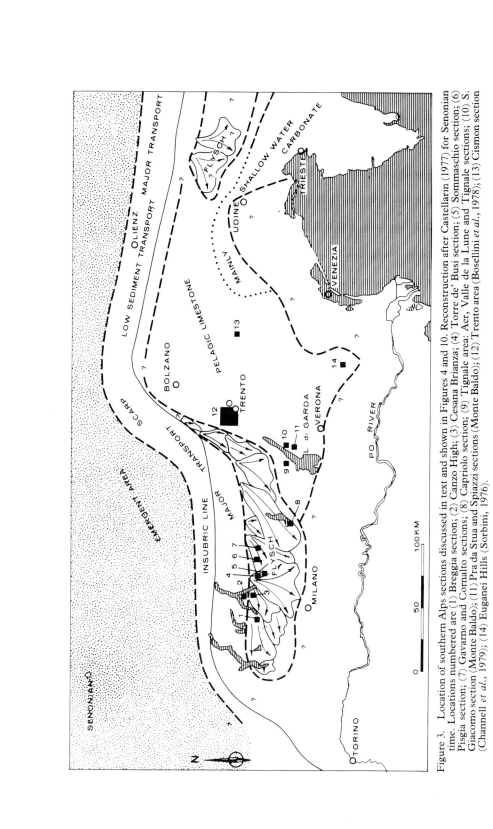

Figure 3.   Location of southern Alps sections discussed in text and shown in Figures 4 and 10. Reconstruction after Castellarin (1977) for Senonian time. Locations numbered are (1) Breggia section; (2) Canzo High; (3) Cesana Brianza; (4) Torre de' Busi section; (5) Sommaschio section; (6) Pisgia section; (7) Gavarno and Cornalto sections; (8) Capriolo section; (9) Tignale area: Aer, Valle de la Lune and Tignale sections; (10) S. Giacomo section (Monte Baldo); (11) Pra da Stua and Spiazzi sections (Monte Baldo); (12) Trento area (Bosellini et al., 1978); (13) Cismon section (Channell et al., 1979); (14) Euganei Hills (Sorbini, 1976).

Cretaceous or Early Tertiary time. Therefore, the region of northern Italy was probably in a semi-arid subtropical zone throughout deposition of most of the Cretaceous pelagic sequence. Upwelling, induced by the prevailing tradewind circulation, may have occurred along the west or north-west margins of the nearby platforms or landmasses.

The sediment in the Umbrian Apennines was deposited above the calcite compensation depth (CCD) level, probably on a continental margin at depths shallower than about 2 km. The depositional setting at Gubbio was isolated from both significant terrigenous input and from redeposited shallow-water carbonate sediment. Elsewhere, turbiditic calcarenitic limestones may occur. Fault-block seamounts may have been present nearby as suggested by Centamore et al. (1971), based on Jurassic facies patterns in Umbria. Condensed pelagic facies occur on the top of these seamounts as at Monte Nerone.

3. Stratigraphic framework

The Cretaceous Umbrian sequences (Figure 4), typified by the Gubbio section, have been subdivided into widely recognizable formations (e.g. Renz, 1951; Merla, 1951; Barnaba, 1958). The Maiolica Limestone is of Late Tithonian through Early Aptian age and consists primarily of gray limestone and nodular chert. The Scisti (or Marne) a Fucoidi are about 50 m thick, of late Early Aptian through Middle Albian age, and consist of varicolored interbedded shale, markstone and limestone. The Scaglia Bianca Limestone comprises relatively thin-bedded gray to creme-colored limestone with thin nodular and bedded cherts of Late Albian through Turonian age (65–70 m thick). The Scaglia Rossa Limestone of Turonian through Eocene age (320–340 m thick) is composed predominantly of pink to red, variably-bedded limestone with nodular chert in the lower part.

Stratigraphic ages are based on the work of Premoli Silva (1977) and Premoli Silva et al. (1977). The stratigraphic scheme for the Southern Alpine and Umbrian sequences is based on Premoli Silva and Paggi (in press). The ages and biostratigraphic correlations, formation names and generalized lithology of Lower Cretaceous sections are shown in Figure 4. Comparison of the Umbrian and southern Alpine sequences is given in the following sections.

4. The Barremian–Aptian Maiolica Limestone and equivalents

4.1. Stratigraphic framework and distribution of facies

The Maiolica Limestone constitutes the lowest Cretaceous sedimentary rock unit and consists primarily of light to dark gray, thick-bedded limestone (av. 25 cm) with interbedded chert layers and thin, sometimes laminated green-gray to black shales. Chert nodules and lenses occur within the limestone beds. The Maiolica Limestone is of Late Tithonian through Neocomian to Early Aptian age (Bortolotti et al., 1970; Lowrie et al., 1980). The Maiolica Limestone and its equivalents (Calcare Ruprestre, Calpionella Limestone) are very widespread in Umbria (Selli, 1952), in the southern limestone Alps (Hsü, 1976; Weissert, 1979) and in parts of Tuscany and Liguria (Figure 1) as a thin and less calcareous calpionellid limestone. The latter probably were deposited on the north-west-facing continental slope of Apulia (Dewey et al., 1973; Channell & Horvath, 1976) and on Upper Jurassic oceanic crust as a more siliceous facies (e.g. Tithonian–Neocomian Calpionellid Limestone and Palombini shales: Neocomian–Albian (?): Elter, 1972; Abbate et al., 1972; Andri & Fanucci,

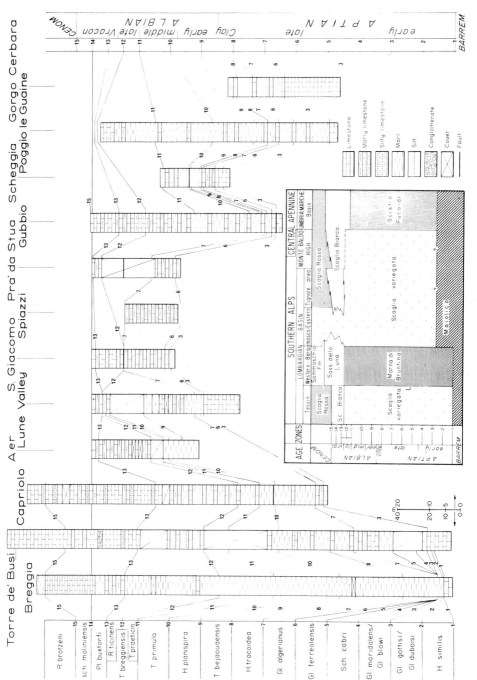

Figure 4.   Biostratigraphy, generalized lithology, and formation names for the Barremian–Cenomanian pelagic sequences of the southern Alps and the Apennines of Italy (after Premoli Silva and Paggi, in press).

1975). During the Neocomian, the CCD was relatively deep and calcareous pelagic sediments were deposited on oceanic crust. This appears to be true of the North Atlantic as well (Thierstein, 1979). The Tuscan region is commonly thought to have been a submarine ridge because of a pronounced disconformity between parts of the Maiolica and overlying deposits of Early Cretaceous to Eocene age (review by Bortolotti et al., 1970). Merla (1951) attributed this disconformity to submarine erosion. In the absence of an angular unconformity, there is no evidence for subaerial erosion or tectonism. The hiatus surface on the Maiolica Limestone is draped by fine-grained pelagic sediment. Bortolotti et al. (1970) have also discussed evidence for submarine dissolution of the upper surface of the Maiolica Limestone, forming a disconformity, and the presence of olistoliths or breccia lenses consisting of fragments of the Maiolica Limestone in parts of the overlying Scisti Policromi (Albian–Cenomanian). They again attribute these relations to the uplift of a submarine ridge between the Umbrian and Tuscan sequence (Figure 1). We propose instead that the Tuscan sequence represents a continental slope facies between the pelagic basin of Umbria and the oceanic facies of Liguria (Figure 2). These relations are now telescoped by the eastward transport and overriding of the Tuscan and Liguride nappes during the Oligo–Miocene (Bortolotti et al., 1970). The disconformities in the Maiolica are hypothesized to have originated by carbonate dissolution and slumping of the carbonate oozes downslope. As previously mentioned, calcarenites (Andri & Fanucci, 1975) and slump or breccia masses (Bortolotti et al., 1970) are found within shales of the Scisti Policromi in Liguria and Tuscany.

The Maiolica Limestone in Umbria is up to about 300 m thick and is composed of Tintinnids (calpionellids), calcareous nannofossils, more rare planktonic foraminifers, ammonites and aptychi with more or less common radiolaria and siliceous sponge spicules seen in thin sections. It appears to be a more shallow equivalent of the facies previously discussed because it is more calcareous and contains more easily dissolved faunal elements, such as ammonites, not found in Tuscan or Ligurian facies (Bortolotti et al., 1970). Centamore et al. (1971) suggested that the Maiolica Limestone draped and smoothed the seamount and basin topography of Umbria which existed during Jurassic times (Figure 2).

The Maiolica in the Southern Alps displays the same characteristics as that from Umbria and Marche regions. The thickness varies from a few meters on topographic highs, such as the Canzo High (north of Milan; Gaetani, pers. commun., 1979), where only the basal part of Maiolica is present, to more than 200 m in the more depressed zones. Slump units and intraformational breccias are intercalated within the Maiolica in several places in basinal or base-of-slope settings. In the eastern part of the Lombardy Basin, thick breccias (i.e. the Breccia di Prabione) occur at the base of the Maiolica, which is younger in age (mainly Middle Berriasian instead of Late Tithonian; Castellarin, 1972; Pasquaré, 1965).

Leaving aside the sections affected by local events, the lower boundary of the Maiolica seems to be synchronous over most of the Southern Alps as it is in Umbria; it is dated as Late Tithonian and coincides with a major change in the pelagic environment (see also Bernoulli, 1972). Calpionellids are the most important pelagic fossils that characterize the interval from Late Tithonian to Valanginian. A marked but gradual decrease in abundance of calpionellids is recorded from bottom to top. This is a generalized feature of the Maiolica all over Tethys (Cita-Sironi, 1964). After the Valanginian, calpionellids occur only very sparsely. After the disappearance of the calpionellids, the main large planktonic organisms in the Maiolica are radiolarians, which vary from abundant to sparse in the calcareous matrix or concentrated in chert nodules and bands. After the Berriasian, the matrix of the Maiolica in which the siliceous zooplankton are suspended is largely composed of nannoconids.

The color of the Maiolica is mostly white; slight-yellow tinges occur at the base of the formation, and the unit becomes gray, and even dark-gray toward the top. Black burrow mottles are very evident in the upper part of the Maiolica, whereas the bottom of the unit appears more homogenous. Bedding ranges from several decimeters in thickness at the base to of the order of one decimeter at most in the upper gray part. Chert, mainly of black to dark gray color, occurs as nodules or bands in the upper part, in association with the increased abundance of radiolaria.

The upper boundary of the Maiolica in the Southern Alps is the same age (Early Aptian) as in Umbria-Marche (e.g. *Globigerinelloides gottisi/Globigerinelloides duboisi* Zone; Premoli Silva & Paggi, in press; Bosellini *et al.*, 1978), except at the Breggia section. Here, the top of the Maiolica is dated, on the basis of ammonites, to the Late Barremian (Rieber, 1977), and planktonic foraminifers are absent. A glauconitic layer, about 30 cm thick, at the top of the Maiolica may represent a condensed section or hiatus and marks the passage to the Scaglia variegata (equivalent to the Scisti a Fucoidi). On submarine highs, the Maiolica appears to have been deposited in more oxygenated conditions. At Monte Baldo (Trento Swell), the Maiolica is not as gray as in the Lombardy Basin or in the Umbria-Marche region, even though it still contains abundant black chert. Bosellini *et al.* (1970) distinguished a slightly pink variety of Maiolica further north on the Trento Plateau, which is either the only Maiolica present (M. Cornetto section) or is only recorded in the upper part of the formation. Chert, in nodules and bands, is always present. The marly interbeds are red and are particularly frequent in the upper part of the formation. Only at the very top of the unit do the beds deposited in the deepest paleoenvironments become more marly and green-gray in color. Long hiatuses occur at the contact with the overlying Scaglia Variegata, which frequently is missing. Sedimentation resumed in such cases during the Late Albian.

4.2. Organic matter in the Upper Maiolica Limestone

Intercalations of black shales up to 10 cm thick in the Upper Maiolica (Barremian and later) are recorded from all the sections surveyed along the edge of the Southern Alps from somewhat deeper paleoenvironments (Weissert *et al.*, 1978). The uppermost layer at the Breggia section, some 3 m below the top of the Maiolica, contains 3.5% Corg. This organic matter displays a high hydrogen and low oxygen index close to that of type 2 (marine) kerogen of Tissot *et al.* (1974). Palynological analysis reveals that the organic matter in this layer is composed of about 60% amorphous matter, less than 10% woody debris and about 35% spores and pollen (Table 1). Large nodules (up to 10 cm diameter) of marcasite and/or pyrite occur in the uppermost part of the Maiolica, near the contact with the overlying varicolored shales, from many localities both in Southern Alps and Umbria-Marche (e.g. at Cesana Brianza, Gorgo a Cerbara, Scheggia, Poggio Morio).

The uppermost 10.5 m of the Maiolica Limestone at Gubbio is shown in Figure 5. Thin (< 5 cm) black shale beds and black cherts interbedded with homogenous to faintly laminated gray limestone are characteristic. The uppermost shale layer, which consists of interbedded black and green-gray intervals, contains 0.71% organic carbon. The limestones typically have less than 0.25% organic carbon. Black shale layers extend back through at least the Barremian and possibly the Hauterivian portion of the Maiolica Limestone and average between 3 and 4% organic carbon; this is also true of the Maiolica Limestone in the southern Alps (Figures 1 and 3; Weissert *et al.*, 1978). These thin, laminated black shales suggest that there may have been intermittent anoxia at the sea floor along the southern continental margin of Tethys.

Table 1A.   Organic carbon content and pyrolysis results from Lower Cretaceous sequences in Italy

Sample	Organic carbon wt (%)	HC Yield (mg/g)	H₂ Index	O₂ Index	T(°C) Maximum pyrolysis	Mean vitrinite reflectance (R₀) %	Age
Moria Bonarelli							
MB-3	2.28	4.055	175	151	427	—	Cenomanian–Turonian
MB-5	2.91	15.100	515	38	419	—	Cenomanian–Turonian
MB-6	6.60	22.654	343	64	426	—	Cenomanian–Turonian
MB-8	2.53	9.616	337	69	420	—	Cenomanian–Turonian
MB-10	16.62	91.761	550	34	419	—	Cenomanian–Turonian
MB-13	15.88	75.206	472	43	425	—	Cenomanian–Turonian
Gubbio Bonarelli							
74713-15J	5.15	23.401	450	45	450	—	Cenomanian–Turonian
74713-15K	3.42	17.786	513	40	513	—	Cenomanian–Turonian
74713-15L	15.66	91.034	574	20	427	—	Cenomanian–Turonian
74713-15M	13.44	81.554	600	24	426	—	Cenomanian–Turonian
*BON-3	17.18	282.2	550	11	420	—	Cenomanian–Turonian
Gubbio Aptian–Albian							
78201	2.66	10.380	389	31	434	0.46	Early Albian
78003	0.15	0.116	56	196	431	0.44	Early Albian
78004	0.30	0.128	31	131	433	0.48	Middle Albian
78005	1.15	1.587	136	64	441	0.42	Middle Albian
*APT-22	7.93	82.5	647	14	421	—	Early Albian
Gubbio Cycle (Albian)							
74710-11F1	0.49	0.133	24	116	440	—	Late Albian
74710-11F2	0.05	0.019	0	920	—	—	Late Albian
74710-11F3	0.48	0.136	22	109	440	—	Late Albian
74710-11F4	0.06	0.014	0	536	—	—	Late Albian
74710-11F5	0.42	0.057	10	132	436	—	Late Albian
Piobbico (Aptian–Albian)							
78101-3	5.21	31.388	600	20	424	—	Early Albian
78102A	1.37	1.521	108	77	445	—	Middle Albian
78102B	2.12	3.994	187	35	437	—	Middle Albian
78103	1.26	1.850	144	49	444	—	Middle Albian
*AER-16	0.30	1.4	16	90	—	—	Middle Albian
*BUSI-58	1.87	22.2	209	30	444	—	Early Albian
*LUNE6	15.06	29.3	595	21	406	—	Late Aptian
*LUNE20	4.17	14.0	399	30	392	—	Late Aptian
*BREGGIA1	3.56	15.6	261	31	420	—	Late Barremian

* Samples marked with asterisk were analysed by IFP; all others analysed in USGS labs (Branch of Oil & Gas Resources).

Table 1B.   Organic carbon content and organic matter type in Lower Cretaceous sequences from Italy

Sample	Organic carbon wt (%)	Amorphous (%)	Type organic matter Spores and pollen (%)	Structured plants (%)	Age
Gubbio					
APT-60	1.12	35	25	40	Middle Albian
APT-34	0.45	25	20	55	Middle Albian
APT-22	7.93	97	3	2	Early Albian
APT-14	0.25	48	12	40	Late Aptian
APT-2	0.96	65	12	23	Late Aptian
Pioggio Le Guaine					
GUA-16	1.02	NC	NC	NC	Middle Albian
GUA-14	1.92	60	22	18	Middle Albian
GUA-8	9.46	80	10	10	Early Albian
GUA-4	0.59	12	18	67	Early Albian
GUA-109A	1.63	75	13	12	Latest Aptian
Scheggia					
SCH-19	1.35	32	28	40	Middle Albian
SCH-17	1.38	35	25	40	Middle Albian
SCH-14A	1.37	30	40	30	Latest Aptian
SCH-1	3.31	75	10	15	Late Aptian
Gorgo Cerbara					
GC-15	3.96	NC	NC	NC	Aptian
GC-14	3.73	75	15	10	Aptian
GC-13A	2.79	37	48	15	Aptian
Lune					
LU-64B	0.48	NC	NC	NC	Vraconian
LY-36	0.22	20	28	52	Late Albian
LU-62	0.39	35	15	50	Middle Albian
LU-31	0.22	NC	NC	NC	Late Aptian
LU-27	0.77	NC	NC	NC	Late Aptian
LU-26	0.13	NC	NC	NC	Late Aptian
LU-20	4.13	50	22	28	Late Aptian
LU-18	0.23	NC	NC	NC	Late Aptian
LU-13	0.24	15	8	77	Late Aptian
LU-6	15.6	NC	NC	NC	Late Aptian
LU-5	3.12	45	25	30	Early Aptian
LU-2	1.12	15	48	45	Early Aptian (?)
AER II					
2AER-27	0.11	NC	NC	NC	Late Albian
2AER-15	0.25	NC	NC	NC	Late Albian
2AER-8	0.28	NC	NC	NC	Late Albian
AER I					
AER-22	0.28	NC	NC	NC	Middle Albian
AER-16	0.24	NC	NC	NC	Middle Albian
AER-13	0.25	5	70	25	Early Albian
AER-11	0.14	NC	NC	NC	Early Albian
AER-5	0.07	NC	NC	NC	Early Albian
San Giacomo					
SG-10	0.34	15	50	35	Late Albian
Capriolo					
C-77	0.66	35	35	30	Late Aptian
C-80	0.63	15	35	50	Late Aptian
C-84	1.28	35	35	30	Late Aptian
Torre De' Busi					
BU-124	0.24	20	30	50	Early Cenomanian
BU-113	0.34	NC	NC	NC	Vraconian
BU-106	0.24	5	15	80	Vraconian
BU-98	0.10	NC	NC	NC	Late Albian

Table 1B. *Continued*

Sample	Organic carbon wt (%)	Amorphous (%)	Type organic matter Spores and pollen (%)	Structured plants (%)	Age
Torre De' Busi *Continued*					
BU-85	2.62	NC	NC	NC	Middle Albian (?)
BU-84	0.49	20	17	63	Middle Albian (?)
BU70B	0.25	NC	NC	NC	Early Albian
BU-134	0.54	10	25	55	Late Aptian
BU-67	0.29	NC	NC	NC	Late Aptian
BU-58	2.10	60	30	10	Early Aptian (?)
BU-55	0.67	NC	NC	NC	Early Aptian
BU-40	0.38	NC	NC	NC	Early Aptian
BU-30	0.76	NC	NC	NC	Early Aptian
BU-29	0.08	NC	NC	NC	Early Aptian
BU-26	0.13	NC	NC	NC	Early Aptian
BU-20	0.08	NC	NC	NC	Early Aptian
BU-17	1.01	88	2	10	Early Aptian
BU-16A	0.31	93	3	4	Early Aptian
BU-7	1.04	15	50	35	Early Aptian
BU-1	7.92	NC	NC	NC	Early Aptian
Breggia					
BR-118	0.15	NC	NC	NC	Late Albian
BR-108	1.35	NC	NC	NC	Late Albian
BR-89	0.48	NC	NC	NC	Late Albian
BR-83	1.01	NC	NC	NC	Middle Albian
BR-77	0.35	6	7	87	Middle Albian
BR-75	1.26	NC	NC	NC	Middle Albian
BR-55	0.41	NC	NC	NC	Early Albian
BR-47	0.20	10	15	75	Early Albian
BR-21	0.79	60	17	23	Late Aptian
BR-6	3.30	58	22	20	Late Aptian
BR-1	3.56	58	36	6	Late Barremian
Cornalto					
M-17	1.28	20	15	65	Cenomanian–Turonian
M-18	0.81	15	15	70	Cenomanian–Turonian
M-21	0.23	35	15	50	Cenomanian–Turonian
Sommaschio					
S-1	0.47	40	30	30	Latest Albian
S-2	0.18	20	15	65	Latest Albian
S-3	0.13	10	20	70	Latest Albian
S-4	0.17	20	50	30	Latest Albian
S-5	5.02	50	20	30	Latest Albian
Gavarno					
G-1	0.42	10	60	30	Cenomanian–Turonian
G-2	1.52	30	40	30	Cenomanian–Turonian
G-3	2.58	15	35	50	Cenomanian–Turonian
G-4	2.59	60	15	25	Cenomanian–Turonian
G-5	0.85	25	25	50	Cenomanian–Turonian
G-6	1.16	15	30	55	Cenomanian–Turonian
G-7	0.61	15	30	55	Cenomanian–Turonian
Pisgia					
P-1	1.21	10	25	65	Cenomanian–Turonian
P-2	2.37	20	30	50	Cenomanian–Turonian
P-3	2.53	20	30	50	Cenomanian–Turonian
P-4	1.99	55	15	30	Cenomanian–Turonian

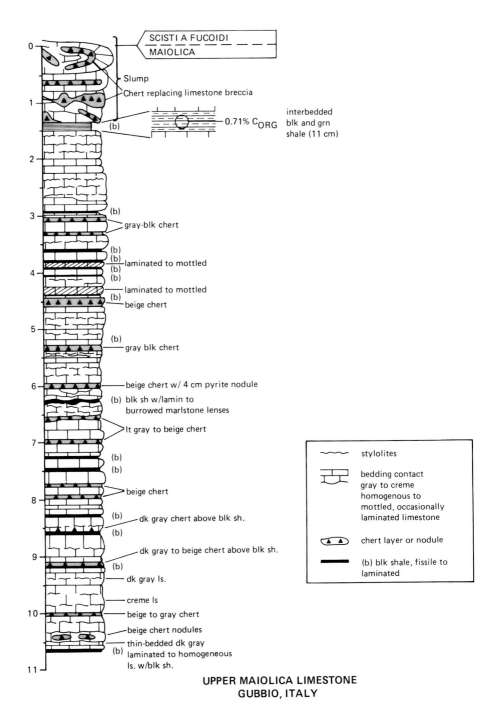

Figure 5.   Uppermost 10.5 m of Maiolica Limestone of probable Late Aptian age at Gubbio, Italy. Section was measured on NW side of road behind old houses and in abandoned quarry. The sequence contains interbedded 25 to 50 cm thick dense gray limestone beds (calcilutites) and thin black and gray fissile to laminated shales. Beige to gray chert up to 10 cm thick occurs in some limestones as beds and nodules. The uppermost 1.5 m of the Maiolica Limestone appears to have been slumped while relatively unconsolidated as evidenced by irregular bedding, folded bedding surfaces and chert layers, and chert-replaced breccias and stretched burrows.

4.3. Contact relations between the Maiolica Limestone and Scisti a Fucoidi

At Gubbio, the uppermost part of the Maiolica Limestone is probably of Early Aptian age (Premoli Silva *et al.*, 1977). The transition from the Maiolica Limestone to the Scisti a Fucoidi above in Umbria is commonly considered to be conformable; there is evidence that there is a disconformity at Gubbio and at two other localities within 20 km of Gubbio. The contact relations at Gubbio, the Valle della Contessa and Piobbico are illustrated in Figures 6 and 7. The uppermost 2–5 m of the Maiolica is apparently slumped at all three localities. The slumping probably took place shortly after deposition because chert-cemented breccias occur in the Maiolica Limestone of the Gubbio section, and because the Scisti a Fucoidi lap over the top with a general conformable nature. However, the lower few meters of the Scisti a Fucoidi is involved in slump folds in the Valle della Contessa. Some later deformation during faulting and folding may have disturbed the contact due to the ductility contrast between the limestone and shale.

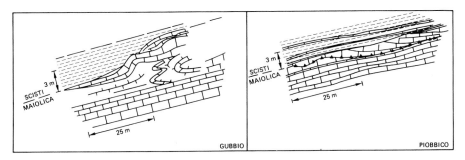

Figure 6.   Sketch of contact relations between the Maiolica Limestone and overlying Scisti a Fucoidi at Gubbio and at Piobbico, about 25 km to the NW. Contact relations suggest a disconformity, possibly due to downslope movement of uppermost portion of the limestone soon after deposition, followed by sedimentation of the shales. Several other lines of evidence suggest a hiatus at the contact, encompassing part of the Early to Late Aptian.

One possible explanation of the slumped uppermost Maiolica is that, contrary to the belief of Centamore *et al.* (1971), relief still existed between nearby seamounts such as at Monte Nerone and the deeper sea floor at Gubbio, even at the end of Maiolica Limestone deposition. The thickness of the Maiolica Limestone in Umbria does vary; at Monte Acuto near Perugia (Figure 1) it is only 130 m thick (Ghelardoni & Maioli, 1959). The CCD level rose abruptly to relatively a shallow depth of about 2 to 2.5 km in the middle to late Aptian (Thierstein, 1979; Figure 8). If the sea floor of the Umbrian basin was about 2 km deep, as earlier suggested, this rise in carbonate solution levels could have caused an increase in dissolution rate in the water column (below the lysocline) by Middle to Late Aptian time. Marlstones and shales of the Scisti a Fucoidi above show evidence of more intense dissolution. Dissolution of pelagic carbonates of the Upper Maiolica Limestone at the base of draped seamounts may have removed support from the seamount slope sediments and caused slumping from the seamount flanks. This mechanism has been suggested to explain slumping of modern pelagic carbonates on the flanks of oceanic rises and seamounts (Berger & Johnson, 1976).

At Gubbio, the lower part of the Scisti a Fucoidi is interbedded with several limestone beds. These beds, which are highly mottled with organic carbon-rich burrows, have abundant closely-spaced microfaults which were apparently initiated while the limestone was still unlithified. The three meters of calcareous shale above are relatively unfossiliferous and contain few age-diagnostic planktonic forminifers

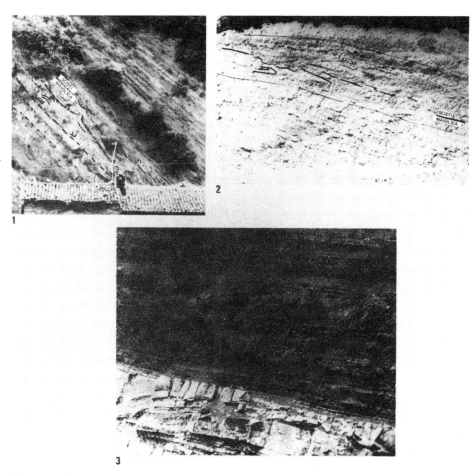

Figure 7.   Contact relations between Maiolica Limestone and Scisti a Fucoidi: (1) Slumped zone in
upper part of Aptian Maiolica Limestone and contact with Scisti a Fucoidi at Gubbio (east wall of
Dola del Bottaccione). (2) Similar slumped zone at contact between Maiolica Limestone and Scisti a
Fucoidi in the Valle della Contessa. (3) Discordant bedding at top of Maiolica Limestone and
overlying horizontally bedded Scisti a Fucoidi at Piobbico, about 20 km NW of Gubbio.

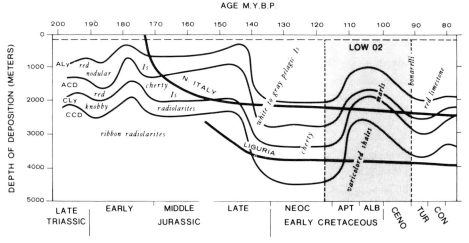

Figure 8.   Inferred subsidence paths (heavy lines) for different crustal elements in Mesozoic Tethys
(note paths of N. Italy—e.g. Gubbio—and Liguria oceanic crust) in framework of carbonate
dissolution horizons (lysocline and CCD) and conditions of oxygenation (modified in part from
Bosellini and Winterer, 1976). CCD levels are same as for N. Atlantic (Thierstein, 1979).

or nannofossils. The maroon fissile shales are enriched in iron relative to the rocks above and below and may indicate very slow deposition under oxidizing conditions. The age of this oxidized zone is Late Aptian (*Globigerinelloides algerianus* zone).

5. Aptian–Albian Scisti a Fucoidi and equivalents—cyclic changes in oxidation state

5.1. Stratigraphic framework and distribution of facies in Umbria

The Scisti or Marne a Fucoidi at Gubbio is about 47 m thick and of probable Late Aptian through Albian age. The unit consists primarily of black to gray shale or marlstone beds alternating with green-gray homogeneous marlstone and limestone beds. Calcium carbonate content ranges from 20 to 80% and generally increases toward the transition to the overlying Scaglia Bianca Limestone (Figure 9). An Upper Aptian oxidized zone (Couches Rouges) having a CaCO$_3$ content as low as 5% occurs in the lower part as previously mentioned. A maximum organic carbon content of 7.93% was measured in a laminated black shale bed. The highest measured organic carbon values occur between 15 and 35 m in the section (Figure 9, Table 1) mainly within the Lower to Lower Middle Albian interval. The planktonic foraminiferal zonation is discussed in the next section along with that of the Lombardy Basin sequences.

The Scisti a Fucoidi is found throughout Umbria (Selli, 1952) and is of equal thickness where measured. A section measured near Piobbico, about 20 km to the north-west, is of the same thickness and shows the same general color and lithologic changes as seen at Gubbio. A section measured in the Marche Apennines near Monte Giove (Colle de Mezzo; roadcuts south of Pieve Torina) about 50 km south-east of Gubbio is again of the same thickness (about 59 m) and succession of lithology as in the Gubbio section. Fazzini and Mantovani (1965) have described the Scisti a Fucoidi as being about 50 m thick near Monte Subasio (near Assisi, see Figure 1).

Similar Aptian–Albian "black shale" facies, usually consisting of varicolored marls and shales, are found throughout northern Italy, in the southern Alps (Belluno Trough, e.g. Channell *et al.*, 1979; Trento Swell; Bitterli, 1965; Lombardy Basin; Venzo, 1954; Gelati & Cascone, 1981 and this paper), in Tuscany as a possible slope facies (lower part of the Scisti Policromi and Broglio shale: e.g. Canuti *et al.*, 1967; Boccaletti & Sagri, 1966), and on Jurassic oceanic crust exposed in Liguria (Palombini shales, Lavagna shales: Abbate & Sagri, 1970). The Lavagna shales are of Albian–Cenomanian age and contain thin turbiditic sands and silts in addition to the gray to black shales and marls. In southern Italy and Sicily, the equivalents are the organic carbon-rich Crete Nere Fm. of Aptian–Albian age (e.g. Ogniben *et al.*, 1975).

5.2. Environment and age of Aptian–Albian sequences from the southern Alps

The Lombardy Basin (Figures 1 and 3) was an elongate east–west feature with an erosional escarpment located on its northern margins. This margin was crossed transversally by a number of canyons and became a location of terrigenous flysch deposition coincident with the first Alpine orogenic phases (Castellarin, 1977; Venzo, 1954; Aubouin *et al.*, 1972; Gelati & Cascone, 1981). The Sass de la Luna Formation, composed mainly of carbonate "pelagic" turbidites with some (rare) coarser intervals (e.g. at Torre de' Busi), is a deposit apparently related to the paleoalpine movements. There is no evidence of tectonic movements during the Cenomanian, but, by the end of the Cenomanian–beginning of the Turonian, the arrival of rapidly deposited flysch marks an important tectonic phase (Gosau).

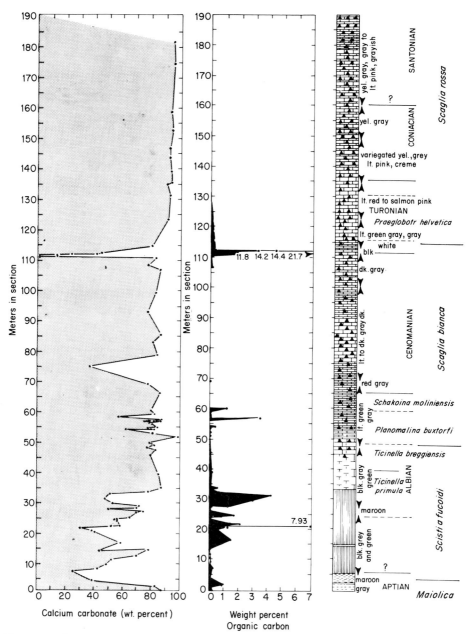

Figure 9.   Generalized lithology, carbonate and organic carbon abundance in the Lower to Middle
     Cretaceous at Gubbio, Italy (data from Arthur, 1979); biostratigraphy after Premoli Silva and Paggi
     (in press).

The sections included in Figure 4 have been chosen because of their different paleoenvironmental setting within the Lombardy Basin (Castellarin, 1976). The Breggia sequence was deposited in a deep basin far from sources of terrigenous debris; only a terrigenous clay fraction is present and pelagic turbidites are rare. The Torre de' Busi section represents deposition in a deep basin, possibly the deepest part of the Lombardy basin, close to some lateral canyons. Terrigenous material and pelagic turbidites are common. A graded conglomeratic layer, almost 1 m thick, occurs in the upper part of the sequence. The Capriolo section was probably deposited on the lower slope or upper continental rise and received some detrital material. Pelagic turbidites are common, particularly in the upper part of the sequence, and large slumps occur in some parts of the section. The Lune/Aer section represents deposition on a fault block flanking the Trento Swell escarpment. There was no detrital clastic supply, but large carbonate slumps were derived from the Trento high, which was under a pelagic regime. Slumps are composed of Maiolica or other Jurassic pelagic sediments, or of black shales of Early Albian age. The S. Giacomo, Pra' da Stua/Spiazzi sequences were deposited on the Trento Swell under a pelagic regime; gaps are marked by faults and a disconformity: the most shaly part, common to other sequences, is missing as it is in the Venetian region (see Cismon section, Channell *et al.*, 1979; Bosellini *et al.*, 1978).

The different settings of the surveyed sections are reflected in the distribution of carbonate content (Figure 10), particularly during the time of deposition of the Scaglia Variegata–Scisti a Fucoidi. In the Breggia and Torre de' Busi sections, the average $CaCO_3$ content in those formations or equivalents is less than 20% (*c.* 14% at Torre de' Busi; *c.* 17% at Breggia). These values reflect averages of all samples analyzed from each sequence which is characterized by rhythmic alternations of relatively $CaCO_3$-rich and poor beds. No attempt was made to sample each bed. Intermediate values between 30% and 40% $CaCO_3$ are typical from slope and medium-depth pelagic plateau sequences (Capriolo in the former, and Aer/Lune the latter). At Gubbio and Poggio le Guaine, the $CaCO_3$ content averages about 50%: these strata possibly were deposited at shallower depths than those in the Aer/Lune sequences. The other sections display medium to high values of $CaCO_3$ (> 50%); however, they have been disregarded because the portion of the Scisti a Fucoidi that is lowest in $CaCO_3$ is missing or not sampled (e.g. at the Gorgo a Cerbara section). It appears that the Apennine sections and the Trento Swell sections are more calcareous than the others belonging to the Lombardy Basin, including the Aer/Lune sections. The uneven distribution of carbonate content through time in a single section is reflected by the patterns of occurrence of planktonic foraminifers in each section. The planktonic foraminiferal zones are as follows.

Globigerinelloides gottisi/G. lloides duboisi Zone: is present in all sections except at Gubbio, where it is missing because of slumping and/or a fault between the Maiolica and Scisti a Fucoidi and at Breggia where the top of the Maiolica is condensed in a glauconitic layer.

G. lloides maridalensis/G. lloides blowi, *Schakoina cabri* and most of *G. lloides ferreolensis* Zones are not documented because of the lack of carbonate, or carbonate content is too low for preservation of planktonic foraminifers (e.g. deposited well below the lysocline). The only exception is the Pra' da Stua section on the Trento Swell.

G. lloides algerianus and *Hedbergella trocoidea* Zones are well documented in all sections. This coincides with the occurrence of the "couches roughes", which are very distinct at Gubbio, Poggio le Guaine, Gorgo a Cerbara, Scheggia, and, to a lesser extent, in all of the southern Alpine sections.

Ticinella bejaouensis and *Hedbergella planispira* Zones are never fully represented. The lower part of the former and the upper part of the latter only are present in some of the sections (Lune, Gubbio, Poggio le Guaine Scheggia). The middle part of this interval consists of the poorest carbonate record in all the sequences. Moreover, the highest Corg values (Figure 10) are recorded in the fish clay (laminated) which

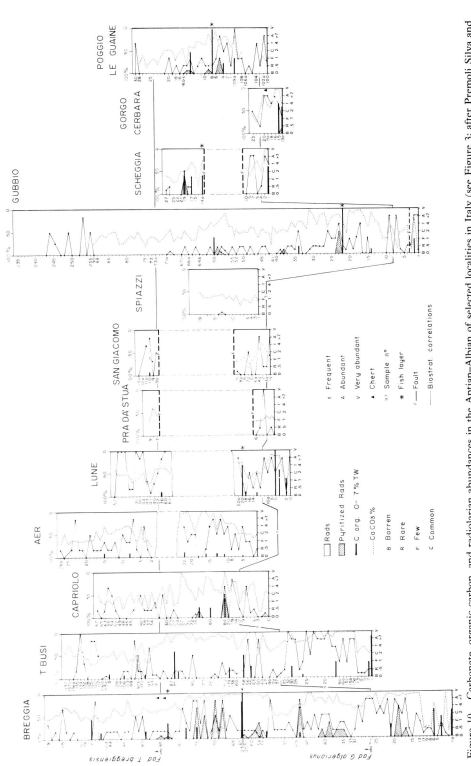

Figure 10. Carbonate, organic carbon, and radiolarian abundances in the Aptian–Albian of selected localities in Italy (see Figure 3; after Premoli Silva and Paggi, in press).

occurs in the middle of this interval. The shale zones are sometimes the slip planes for faults (Spiazzi, S. Giacomo, Scheggia).

Ticinella primula Zone is recorded in all sections except on those from Monte Baldo (Trento Swell) where it apparently is cut out by a fault. Foraminiferal faunas, both planktonic and benthic, become richer in this zone, even though not yet well diversified.

Ticinella breggiensis Zone as well as the *Planomalina buxtorfi* Zone, the youngest considered here, are recorded from all the sections in which this interval was sampled.

5.3. Benthic assemblages from Aptian–Albian sequences

Benthic foraminifers occur throughout all of the sequences studied, except that their abundance and diversity vary from layer to layer. Benthic assemblages, particularly those forms with calcareous tests, appear to be affected by dissolution and by degree of tolerance to poorly-oxygenated conditions. Simple agglutinated forms, such as *Rhizammina*, *Bathysiphon* and *Haplophragmoides* are either solution-resistant species or adapted to poorly-oxygenated environments. This fauna, characterizing the darker, less calcareous layers, can be identified as a "residual fauna", undiagnostic for depth evaluation.

More complete benthic assemblages from limestone layers contain a relatively large number of agglutinated species and specimens, such as those mentioned above, associated with finely textured *Gaudryina*, *Bigenerina*, etc., and calcareous forms such as *Gavellinella*, *Gyroidina*, *Osangularia*, *Pullenia*, *Pleurostomella*, *Stilostomella* and a few nodosariids. This assemblage is indicative of a lower bathyal to abyssal environment (Sliter, 1977). The assemblage occurs, with almost no variation, in all of the surveyed sections except those from the Monte Baldo area (Trento Swell). In the Spiazzi, S. Giacomo and Pra' da Stua sections, nodosariids are more abundant, but most of the upper bathyal forms are missing. Thus, on the basis of benthic faunas, the environment at those localities was primarily middle bathyal (upper bathyal— 200–500 m; middle bathyal—500–1500 m; lower bathyal—1500–2500 m; abyssal >2500 m).

5.4. Lithology and paleontology of the Scisti a Fucoidi at Gubbio: an example. The
Scisti a Fucoidi is characterized by cyclic alternations of black shale or marlstone and greenish-gray, more homogeneous marly limestone and limestone (Figures 11 and 12). A representative cycle is shown in Figure 13. It is very similar to, but generally more carbonate-rich, than cycles from the Aptian–Albian of the North Atlantic cored in DSDP sites (Dean *et al.*, 1978; McCave, 1979). At Gubbio, the black to gray calcareous shales and marlstones at the base of each cycle are homogeneous to laminated and are of variable thickness, generally less than 10 cm. Contacts between these and the more massive, homogeneous green-gray marly limestones and limestones are transitional to sharp and often burrowed. Burrows of dark-colored sediment rarely penetrate more than 5 cm into the green-gray limestone below. The primary trace fossil is *Chondrites*, but *Planolites* traces commonly occur in a number of beds in the upper part of the Scisti a Fucoidi. Most burrow systems observed in outcrop are shallow and typically along beds rather than penetrating deeper into sediment rich in organic carbon.

Pyrite is common to abundant, mainly in the green-gray limestone beds. Organic carbon content is usually higher in the laminated, black, relatively carbonate-poor intervals than in the homogeneous, more calcareous upper part of each cycle. A

Figure 11. Lithology of the Scisti a Fucoidi. (1) Laminated "black shales" of lower part of an oxidation–reduction cycle. Sample from this level contained 6.37% organic carbon. Large pyrite nodules occur in this bed (top middle). Scale is 10 cm long. (2) Bedding plane on upper surface of a light green-gray limestone in upper part of formation showing *Chrondrites* and *Planolites. Chrondrites* is the most common trace fossil. (3) Upper part of the black shale/light green-gray marlstone cycle. Characteristic *Chrondrites* traces are filled with organic carbon-rich material.

number of such cycles from the base to the top of the Scisti a Fucoidi at Gubbio are shown in Figure 14.

The relatively high organic carbon content, often well-preserved thin laminations, and the general lack of burrowing in the dark-colored portion of cycle suggest that anoxic or near-anoxic conditions existed on the sea floor during deposition of this interval. The burrowed transitions from more homogeneous green-gray limestone to black, laminated marly shale suggest that anoxic conditions were typically established gradually. Hence the sequence of lithology usually goes from massive green-gray limestone through a more fissile greenish-gray limestone with black *Chondrites* burrow traces to more homogeneous, splintery streaked black to gray

Figure 12. Bedding and oxidation–reduction cycles in the Scisti a Fucoidi. (1) Middle part of Scisti a Fucoidi at Piobbico, Italy, about 20 km NW of Gubbio. Stratigraphy is very similar to that at Gubbio. Black shale/gray marlstone cycles are obvious here. (2) Transition between the Scisti a Fucoidi (Aptian–Albian) and the Scaglia Bianca Limestone at Piobbico. Black shale/gray-green marlstone cycles gradually give way to maroon or reddish-gray limestone with interbedded fissile marlstone (Middle to Upper Albian), then to creme-gray limestone of the Scaglia Bianca (Upper Albian). (3) Scisti a Fucoidi: Black shale—green fissile marlstone—green-gray limestone.

calcareous shale, and then commonly to thinly-laminated, papery black shale. The thickness of these intervals varies, and, in some cases, there is no laminated interval. The upper part of the black interval is also commonly homogeneous to burrow mottled. It is not clear whether or not oxygenated conditions were resumed rapidly since the black portion of the cycle has obviously been burrowed during deposition of the green-gray limestone and the original contact relations destroyed. The green-gray limestone beds were probably deposited under mildly oxygenated conditions because burrowing occurs in these intervals, but reducing conditions were established shortly after deposition as indicated by authigenic pyrite and by the fact that organisms did not burrow the sediment through more than one cycle at a time.

Samples of each type of lithology have been disaggregated, washed, and sieved (> 62 μm) to examine the composition of the coarse fraction. This was done in the

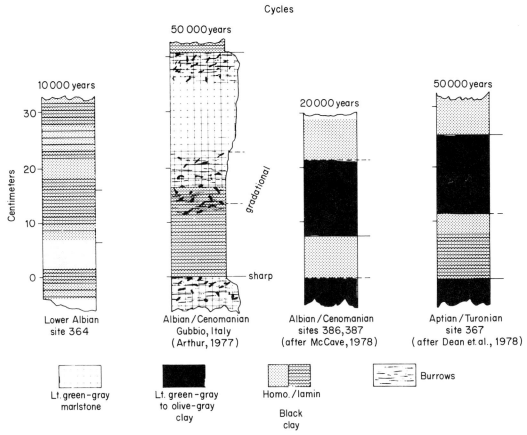

Figure 13. Representative carbonate-organic carbon cycles from Lower to Middle Cretaceous sequences in Tethys and the North Atlantic (after Arthur and Natland, 1979).

less carbonate-rich interval in the Scisti a Fucoidi because of the ease of disaggregation and in order to preserve any calcareous elements. Thin sections of many samples have also been examined, but it is often difficult to obtain a good section of the black-shale intervals.

Carefully selected and washed samples of completely laminated black shale contain primarily fine silt-sized angular quartz grains, some pyrite, phosphatic material (Ichthyoliths), occasionally a few arenaceous benthic foraminifers, and some poorly-preserved radiolaria.

More homogeneous, burrowed intervals of gray to olive or black shale ($>10\%$ $CaCO_3$) contain rare calcareous plankton, sometimes pyritized, and also sometimes contain an impoverished calcareous and/or arenaceous benthic foraminiferal assemblage.

The washed residue of samples of homogeneous green-gray limestone contain abundant but relatively poorly-preserved planktonic foraminifers, calcispheres, a few calcitized and/or pyritized radiolaria, common pyrite, ostracod valves, echinoid (?) spines or sponge spicules. some phosphatic debris, and an abundant and diverse calcareous benthic foraminiferal assemblage similar to that described from the southern Alpine sections.

In thin section, the laminated black shale samples (Figure 15) show alternating thin discontinuous laminae (<0.1 mm thick) consisting of wisps or flakes of brown

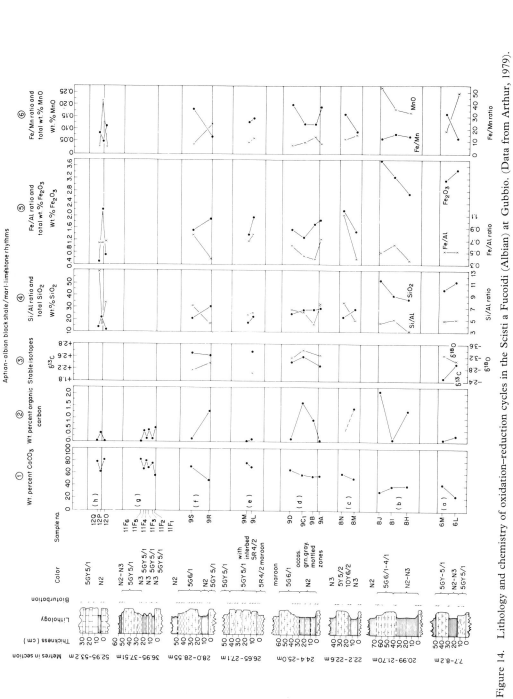

Figure 14. Lithology and chemistry of oxidation–reduction cycles in the Scisti a Fucoidi (Albian) at Gubbio. (Data from Arthur, 1979).

organic matter and partly silicified micritic calciate. Platy phosphatic detritus (<0.01 mm diameter) is also common. Rare, very poorly-preserved planktonic foraminifers also occur. The tests are usually filled with sparry calcite. More or less randomly-oriented single ostracod valves (?) (0.05–0.20 mm long) and echinoid (?) spines or possibly calcitized sponge spicules, large calcareous and arenaceous benthic foraminifers, and rare calcispheres have also been observed. Common silt-sized quartz (3–5%, <0.02 mm diameter) occurs in many sections. Pyrite grains and phosphatic debris also occur. The limestones generally have a homogeneous micritic fabric, but faint laminations are sometimes present; planktonic foraminifers are much more abundant than in the black shale intervals.

The evidence summarized above indicates that periodic changes in the supply of organic matter and/or extent of oxygenation occurred on or above the sea floor. These changes appear to be cyclic with a period near 20 000–50 000 years (Arthur, 1979a).

5.5. Origin of Aptian–Albian oxidation/reduction and carbonate cycles

Weissert et al. (1978) have suggested, on the basis of lithology and $\delta^{13}C$ values of carbonates, that black-shale limestone cycles in the Barremian of the Maiolica Limestone represent periodic changes from a stably stratified anoxic basin to a more oxygenated overturning basin. Similar cycles have been described from Lower Cretaceous sequences in the North Atlantic (Dean et al., 1978; McCave, 1979; Arthur, 1979a) and have been ascribed to varying causes including stagnation and anoxia in deep waters above the sea floor, periodic changes in oxidation of sediment at the sea floor in the presence of oxygenated waters above the sediment/water interface, and as productivity cycles. A further problem is whether the organic matter in the "black shale" cycles is terrigenous or marine, as this bears on the question of sources of organic matter of possible enhanced preservation of organic matter and causes of anoxia.

There is presently no adequate method of detecting productivity variations in ancient sediments because of the lack of precise time control for calculating accumulation rates, because we do not know the superimposed effects of dissolution in the water column and on the sea floor, and because of diagenetic changes in the sediment. However, some speculations can be made based on lithologic, geochemical and faunal associations, and bulk accumulation rates.

It appears that most of the Scisti a Fucoidi in the Umbrian region were deposited at relatively slow rates. The figures cited here are based on the van Hinte (1976) time scale and may be in error by a factor of two. The interval from 5 to 35 m is characterized by generally low calcium carbonate contents (Figure 9). The interval from 5 to 25 m represents the Lower and Middle Albian and the uppermost part of the Aptian as well (Figure 9). The Ticinella breggiensis zone at 25 m is the base of the Upper Albian (Premoli Silva et al., 1977). Therefore, the carbonate-poor interval, which is the main part of the sequence characterized by the oxidation/reduction cycles, was deposited at rates of less than 5 m per m.y. (0.5 cm per 10^3 years). This is half the rate calculated for the Upper Albian part of the sequence where $CaCO_3$ contents increase rapidly and chert nodules and lenses become more abundant, and about the same as the rate of deposition of the Cenomanian and Turonian cherty limestone sequence. The accumulation rate of aluminum, calculated from the averages from chemical analyses and bulk density determinations by Arthur (1979b) and using sedimentation rates of Premoli Silva et al. (1977), decrease from 0.058 g cm^{-2} per 10^3 years in the Lower to Middle Albian, to 0.034 g cm^{-2} per 10^3 years in the Upper Albian (Vraconian) and to 0.010 and 0.011 in the Cenomanian and Turonian, respectively. Fine silt-sized angular quartz grains are common in the lower Scisti a

Fucoidi. The calculated accumulation rate of $CaCO_3$ is only 0.70 g cm^{-2} per 10^3 years in the Lower to Middle Albian, whereas rates increase to 2.25 g cm^{-2} per 10^3 years in the Vraconian, and to 0.99 to 1.24 g cm^{-2} per 10^3 years in the Cenomanian and Turonian. It appears that the terrigenous influx could have been up to five times greater during the deposition of the $CaCO_3$ poor part of the Scisti a Fucoidi than during the remainder of Albian through Turonian; while the $CaCO_3$ flux to the sea floor was nearly 50% lower. These are only estimates, and their accuracy depends mainly on the accuracy of the time scale used (van Hinte, 1976; Sigal, 1977).

The lower rates of accumulation of $CaCO_3$ suggest a low carbonate supply during the Late Aptian and Lower to Middle Albian due either to high rates of dissolution or to low surface productivity. The CCD in the North Atlantic and Pacific oceans rose during Aptian time and remained at about 2–2.5 km depth until Late Albian or Cenomanian time when it dropped slightly (Thierstein, 1979). The general decrease in carbonate supply in the uppermost Aptian and Lower to Middle Albian at Gubbio may be a response to the rising CCD and probably a rising lysocline (Figure 8). Planktonic foraminifers are rare and more fragmented in the Lower Albian Scisti a Fucoidi than in the Upper Albian, and this may be a sign of universal increase in dissolution at this time (e.g. Sliter *et al.*, 1975).

Variations in the $CaCO_3$ content from all of the studied sections are similar to that at Gubbio. The Tithonian–Lower Aptian section is characterized by high $CaCO_3$ content, up to 90%; the middle part of the Aptian is characterized by a general drop in carbonate content with values rarely exceeding 40% and frequently by beds with no carbonates at all; the Upper Aptian records short excursion to higher values of $CaCO_3$ of about 50% followed by an interval with very poor carbonate preservation, the latter spanning the Upper Aptian and the Lower Albian. In the Middle Albian section, the carbonate content is high ($>70\%$) in the transition from the Scisti a Fucoidi/Scaglia Variegata to the Scaglia Bianca. The Upper Albian strata are generally very rich in carbonate, but the deep basinal sections from the Lombardy Basin are less calcareous than those from the Umbria region.

The bulk silica/alumina ratio also increases from Aptian through Cenomanian and Turonian strata. A value of the Si/Al ratio of about three is considered to be an average ratio for terrigenous sediment (Garrels & MacKenzie, 1971; W. Donnelly, pers. commun.). Values of greater than three are generally thought to be due to an "excess" silica supply, probably from a biogenic component. Although total silica decreases upward in the section, Si/Al values are relatively low (3–7) in the Lower to Middle Albian and generally increase to over 9 in parts of the Upper Albian and Cenomanian–Turonian; thus, the proportion of biogenic silica apparently increases. This increase corresponds also to a gradual increase in the number and size of chert nodules from the Upper Albian through the Turonian. The first Upper Albian chert nodules follow about 10 m above an increase to higher $CaCO_3$ values (over 80%). The excess silica probably originated as radiolarians which now occur as ghosts in chert nodules or commonly as poorly-preserved specimens in the Scisti a Fucoidi. The chemical and lithological evidence suggests a general trend toward increasing productivity or preservation of silica and $CaCO_3$ from Early Albian through Turonian time. If this is correct, then the black shales within the Scisti a Fucoidi, which are rich in organic carbon and low in $CaCO_3$ probably neither resulted from preservation under generally high sedimentation rates, as is commonly argued (Scientific Staff, DSDP Leg 48, 1976; Habib, 1979; Müller & Suess, 1979), nor under continuous exceptionally high productivity conditions as might be expected. Distribution of radiolarians in the time-equivalent sections from the southern Alps is plotted in Figure 10. Even though the plotting of samples is not to scale, the biostratigraphic tie-points should help to understand the radiolarian distribution. Radiolarians are present throughout all sections. Below the *Ticinella primula* Zone

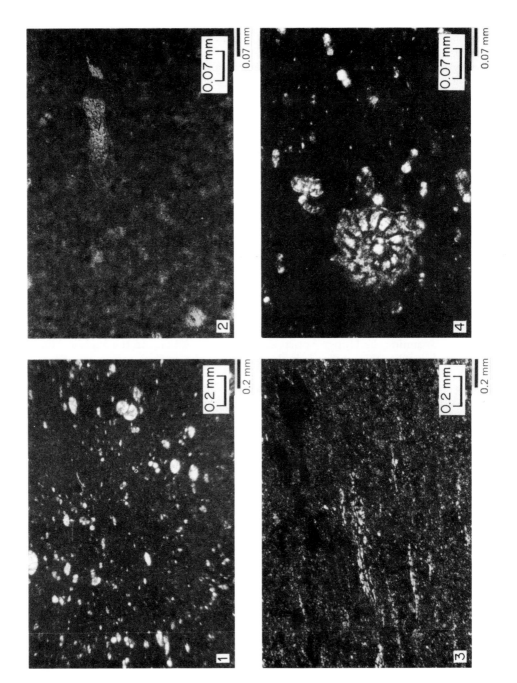

0.07 mm

0.07 mm

0.07 mm

0.07 mm

0.2 mm

0.2 mm

0.2 mm

0.2 mm

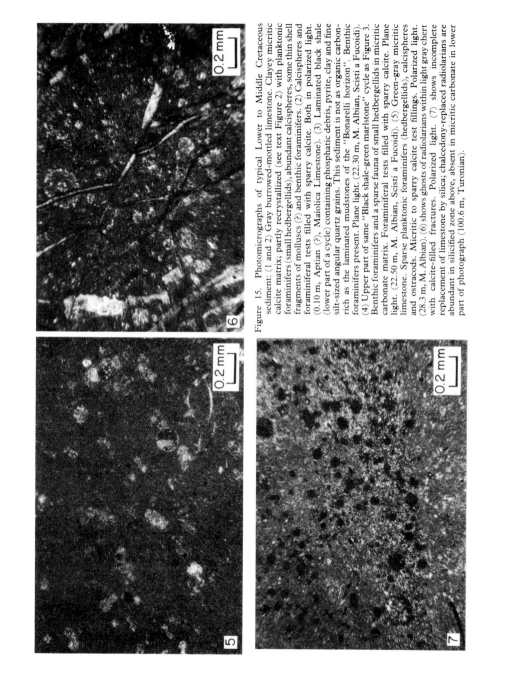

Figure 15. Photomicrographs of typical Lower to Middle Cretaceous sediment: (1 and 2) Gray burrowed-mottled limestone. Clayey micritic calcite matrix; partly recrystallized (see text Figure 2) with planktonic foraminifers (small hedbergellids), abundant calcispheres, some thin shell fragments of molluscs (?) and benthic foraminifers. (2) Calcispheres and foraminiferal tests filled with sparry calcite. Both in polarized light. (0.10 m, Aptian (?), Maiolica Limestone). (3) Laminated black shale (lower part of a cycle) containing phosphatic debris, pyrite, clay and fine silt-sized angular quartz grains. This sediment is not as organic carbon-rich as the laminated mudstones of the "Bonarelli horizon". Benthic foraminifers present. Plane light. (22.30 m, M. Albian, Scisti a Fucoidi). (4) Upper part of same "Black shale-green marlstone" cycle as Figure 3. Benthic foraminifers and a sparse fauna of small hedbergellids in micritic carbonate matrix. Foraminiferal tests filled with sparry calcite. Plane light. (22.50 m, M. Albian, Scisti a Fucoidi). (5) Green–gray micritic limestone. Sparse planktonic foraminifers (hedbergellids), calcispheres and ostracods. Micritic to sparry calcite test fillings. Polarized light. (28.3 m, M. Albian). (6) shows ghosts of radiolarians within light gray chert with calcite-filled fractures. Polarized light. (7) shows incomplete replacement of limestone by silica; chalcedony-replaced radiolarians are abundant in silicified zone above, absent in micritic carbonate in lower part of photograph (100.6 m, Turonian).

(Middle Albian), radiolarians are mainly concentrated in layers comparable to radiolarian sandstone which are evidenced in outcrop because they are more resistant to weathering than the interbedded shales and/or marls. Radiolarians are also the largest component of pyritic claystone, where they represent almost 100% of the $>63\ \mu$m fraction: in those layers, radiolarians are fully pyritized.

Above the *T. primula* Zone, radiolarians occur first in silicified limestone and then almost exclusively in chert nodules or bands. Chert is, in general, associated with highly calcareous sediments; while it appears that abundant clay prevented chertification in the lower part of the Scisti a Fucoidi. In the Scaglia Variegata or Scisti a Fucoidi, some quartz is present mainly as authigenic crystals. These are particularly frequent in the Gubbio section.

Geochemical analyses have been performed on a few samples from the Lune and Torre de' Busi sections in addition to the Gubbio section. The total weight percent SiO_2 and the Si/Al ratio are listed in Table 2. In all samples, the Si/Al ratio is higher than three; however, at Torre de' Busi, the average ratio is lower than in the other two sections; that is consistent with the occurrence of more detrital material in the strata from Torre de' Busi (see above). Note that sample APT22, the richest in Corg content from the Gubbio section in this interval, has a Si/Al ratio close to 13, suggesting that most of the silica is biogenic in origin. A high ratio (*c.* 8) also occurs in sample LU20, which again is very rich in Corg. It seems that some of the laminated black-shale layers of Early Albian age are high in biogenic silica and mark the maximum of accumulation of Corg of marine origin in the Early Cretaceous of the Mediterranean Tethys region (see below). This suggests that although productivity was not high during the Aptian–Early Albian, some individual laminated black

Table 2. SiO_2 and Si/Al of selected black shales from the Gubbio, Torre de' Busi and Lune Sections, Italy

Sample	SiO_2 weight (%)	Si/Al*	Age
Gubbio			
APT9	37	10.2	Late Aptian
APT12	34	4	Late Aptian
APT17	51	4.2	Early Albian
APT22	45	12.8	Early Albian
APT29	37	6	Early Albian
APT56	38	5.8	Middle Albian
APT60	27	4	Middle Albian
APT77	29	4.2	Late Albian
APT79 (red)	34	5.1	Late Albian
APT79 (grn)	28	4.8	Late Albian
Torre de' Busi			
BU138	55	3.8	Early Albian
BU70A	52	4.1	Early Albian
BU95	23	5.3	Late Albian
BU102	27	6.8	Late Albian
BU108A	40	4.6	Vraconian
BU116	23	7.5	Vraconian
BU137A	33	4.2	Early Cenomanian
BU137	32	3.7	Early Cenomanian
Lune			
LU8	58	5	Late Aptian
LU11	35	4.3	Late Aptian
LU20	48	8	Early Albian
LUH1	10	3.7	Early Albian
LUH2	22	4.1	Early Albian

* Determined by Atomic Absorption (R.S.M.A.S., U. of Miami and U. Milano; courtesy of E. Bonatti and Tiberto; *not* expressed on $CaCO_3$-free basis).

shales may record periodic high productivity pulses. Both samples APT22 and LU20 also contain a large amount of fish debris.

The sporadic occurrence of such beds and the concentration of radiolarians in discrete layers might also suggest redeposition of organic carbon-rich mud from adjacent submarine slopes. It is difficult at this time to demonstrate that such beds are not turbiditic in origin. At least some of them have sharp basal contacts, contain detrital silt and clay minerals, and exhibit lamination atypical of varved sediments. No evidence of graded bedding was observed in any of these layers, but this is not negative evidence since turbiditic muds are often not graded (e.g. Hesse, 1975).

Organic carbon-rich strata of equivalent age in the North Atlantic contain an appreciable fraction of organic matter apparently derived from terrigenous higher plants (Tissot et al., 1979, 1980; Summerhayes, 1981) even though the sedimentary structures in many layers suggest that anoxic conditions probably existed at or above the sea floor during deposition. Relatively little organic matter of marine derivation is present, even though conditions were apparently right for preservation. The most notable exception to this general trend is the region under a probable Early Cretaceous upwelling zone off N.W. Africa where high Corg contents and marine amorphous Corg predominate (Tissot et al., 1979, 1980; Summerhayes, 1981). Geochemical and optical studies of the organic fractions provide data on the amount, type, and oxidation state of organic matter in the black shales. Optical studies (provided by R. Farraris and G. Piccaia, University of Milan) suggest that a small fraction (5–20%) of the organic matter in some black-shale intervals consists of cuticles, spores, and other structured material derived from terrestrial higher plants. However, much of the material is amorphous and may be of marine origin. The highest Corg values measured in the Scisti a Fucoidi was 7.98% at Gubbio, which is high for a pelagic sediment (McIver, 1975), but not unusual for the Lower Cretaceous. Pyrolysis studies of organic matter (Figure 16) (provided by G. Deroo, Institute France du Petrole and G. E. Claypool, U.S.G.S. Denver) suggest that the higher amounts of organic carbon in some beds are due to preservation of marine organic matter having relatively high hydrogen and low oxygen indices. However, in black-shale layers having organic carbon contents below about 1%, the organic matter has a lower hydrogen and higher oxygen index. This suggests either a higher level of oxidation of the sedimented organic matter because of deposition under more oxygenated conditions or a relatively larger proportion of terrigenous organic matter. We suspect that a combination of the two effects is responsible.

Corg has been measured from darker layers in sections from the Southern Alps. The richest layers are black to very dark gray in color; their Corg content ranges from less than 1% Corg to 13.6% in the Scisti Ittiolitici at the Breggia (Bitterli, 1965) and 15.6% in Lune section, sample LU6. The Corg values obtained are listed in Tables 1A and 1B. The other lithologies such as dark olive-green claystone, do not contain Corg in an appreciable amount; whereas, the red marls are devoid of organic matter.

In all the sections, the highest Corg values occur in the Early Albian interval between the middle T. bejaouensis Zone and the lower H. planispira Zone. In this interval, as mentioned above, planktonic foraminifers and calcareous nannoplankton are very rare or absent.

There is a positive correlation between the total amount of organic matter and the amount of amorphous organic matter as determined by microscopic examination of samples under reflected light. Pyrolysis analyses on some of the samples show that the kerogen in strata with high Corg content has high hydrogen indices and is probably of marine origin (e.g. Tissot et al., 1974). The opposite occurs in samples containing a low amount of organic carbon: terrestrial plant material, characterized by a high oxygen and low hydrogen index, becomes the dominant component of the organic matter. Consequently, we interpret the amorphous organic matter to be

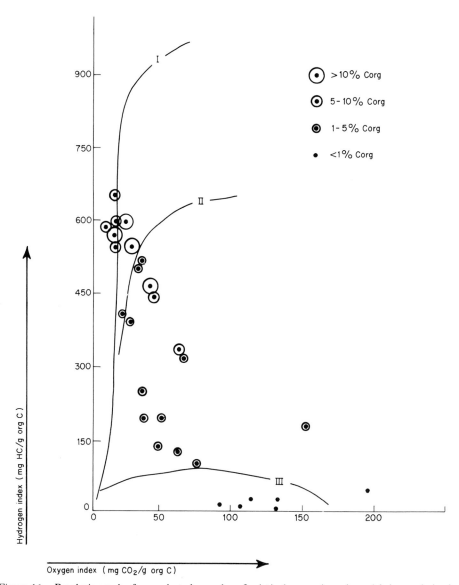

Figure 16. Pyrolysis results from selected samples of relatively organic carbon-rich intervals in the Lower to Middle Cretaceous pelagic sequences in the Umbrian and S. Alpine region. The size of the circle is proportional to organic carbon content with four sizes smaller to larger representing 1%, 1–5%, 5–10%, and 10% Corg. Clear circles represent samples from the "Bonarelli" horizon and equivalents and dark circles samples from the Scisti a Fucoidi. Symbols I, II, and III are the organic matter classifications of Tissot et al. (1974) discussed in text.

primarily marine in origin. The input of terrestrial Corg appears to have been constant in all of the measured sections but slightly higher in the sections from the Lombardy Basin. This is consistent with the location of the Lombardy region much closer to land areas under erosion at the time.

The deepest part of the Lombardy Basin contains a second episode of accumulation of Corg before the end of the Albian, mostly during the *P. buxtorfi* Zone. More than 10 m of black shales with up to 5% Corg occur in the Sommaschio section just south of the Torre de' Busi section; two layers, 10 cm thick, containing 1.35% and

0.15% Corg, respectively, have been found at the Breggia section in the same foraminiferal zone. About one meter of black shales, rich in uranium are recorded at Mollaro and Aldeno, north-west of Trento on the Trento Swell (Fuganti, 1966; Bosellini et al., 1978).

In shallower, much more carbonate-rich sections, this episode appears as a layer less rich in carbonate (e.g. at Gubbio, S. Giacomo).

The deposits at Gubbio, relatively rich in organic carbon, must represent periodic oxygen depletion in deeper water coupled with changes in the production of organic matter in surface waters. Laminated organic carbon-rich black shale portions of the oxidation–reduction cycles, which have a period of from 20 000 to 50 000 years, may actually represent times of reduced productivity during enhanced stability of the water column. The carbonaceous shales or marlstones in each cycle are thinner, have higher contents of organic carbon, lower $CaCO_3$ content, and, often, but not always, lower Si/Al ratios (Figure 14) than the upper burrowed limestone portion of each cycle. The limestones, although initially deposited in a mildly oxidizing basin, have lower MnO_2 contents on a carbonate free basis and higher Fe/Al and Fe/Mn ratios, probably because of the more extensive formation of authigenic pyrite in anoxic pore waters after deposition. Stable carbon isotope analyses done on samples across several cycles suggest that bulk carbonate in the organic carbon-rich intervals (Figure 14) is isotopically heavier than that in the green-gray limestones, a relationship similar to that found by Weissert et al. (1978). If this is a general relationship, it could indicate short-term stable stratification and depletion of ^{12}C in surface waters during deposition of the laminated shales, and higher rates of upwelling and replenishment of ^{12}C in surface waters during deposition of the limestone beds. However, we cannot state with certainty that the more organic-rich layers are not due to greater upwelling and production of organic matter.

The Aptian–Albian Scisti a Fucoidi of Gubbio and Umbria region are similar to, and age equivalents of, widespread carbonaceous sediments deposited at virtually all depths in the North and South Atlantic and in Tethys. Although relatively rich in organic carbon, these deposits may not have resulted from a period of increased production of organic matter, but probably resulted from a combination of episodic productivity and depletion of oxygen in deeper water masses. Evidence from the Pacific Ocean basin suggests that Aptian-Albian organic carbon-rich sediments resulted from periodic oxygen-deficient conditions due to expansion of an oxygen-minimum layer (Schlanger & Jenkyns, 1976; Fischer & Arthur, 1977) possibly modified by local structural and hydrographic conditions around mid-ocean equatorial seamounts and plateaus (Thiede et al., this volume). Sluggish bottom-water circulation may have resulted from relatively long-term stable stratification from high salinity contrasts between surface and deep waters in the South and possibly North Atlantic (Ryan & Cita, 1977; Arthur & Natland, 1979; Thierstein & Berger, 1978; Thierstein, 1979). Bottom waters of high salinity may have been derived by spillage either from the isolated evaporitic South Atlantic (Angola Basin) or the widespread low-latitude evaporitic shelves (e.g. Arthur, 1979b). The inferred warm temperatures of the intermediate and deep water masses (e.g. Savin, 1977) probably also led to decreased oxygen solubility and development of more widespread oxygen deficits. Organic matter may also have been redeposited from the outer shelves and slopes to deeper basins (e.g. de Graciansky et al., 1979; Habib, 1979; Dean & Gardner, this volume).

The Umbrian sequence was probably deposited in water less than 2 km deep, so it is clear that oxygen-deficient conditions extended from deeper Tethyan ocean basins (Liguria: Palombini and Lavagna Shales) to shallower pelagic basins of plateaus including the Trento Swell (e.g. Bitterli, 1965). Carbonaceous Orbitolina marls of probable Aptian age are even found interbedded in the platform carbonate facies of Abruzzi (Figure 2; d'Argenio, pers. commun.; Praturlon & Sirna, 1976).

6. The "Bonarelli" horizon and the Cenomanian/Turonian "anoxic event"

6.1. Cenomanian–Turonian "anoxic event"

The "Livello Bonarelli" (Bonarelli horizon) or Scisti Ittiolitici was recognized as an unusual bed in the Gubbio sequence (Figure 17) long ago by Renz (1936, 1951) and Barnaba (1958). This less than 1 meter thick bed consists of radiolarian silts and

Figure 17. Early Turonian "Bonarelli horizon": (1) "Bonarelli horizon" at Madonna del Sasso (cemetery) about 1 km SW of the Gola del Bottaccione and Gubbio section. This 1 m thick phosphatic, radiolarian interval containing organic carbon occurs at 111 m in the Gubbio section. The bed occurs at many localities throughout Umbria. (2) "Bonarelli horizon" at Cave de Cecchetti, about 5 km NW of Gubbio. This layer occurs within a sequence of gray to creme cherty limestones which change gradually to red and pink limestones about 15 m above the bed (upper right of photo). (3) Lithology of the "Bonarelli horizon" at Moria, about 20 km N of Gubbio. Interval is about 1 m thick (scale = 20 cm). Resistant light-colored beds are radiolarian silts. Dark intervals are laminated organic carbon-rich mudstone with abundant phosphatic debris and "excess" silica. (4) Sample 74713-15N1 from upper part of "Bonarelli horizon", laminated, with lenses of radiolarians in upper part, and rare burrow mottles. Typical of black mudstones in this interval.

interbedded phosphatic mudstone, rich in organic carbon, is widespread in the Umbria region and has commonly been used to mark the boundary between the Scaglia Bianca Fm. and the Scaglia Rossa Fm. (Figure 9; Selli, 1952). The Bonarelli horizon spans the Cenomanian–Turonian boundary in the Gubbio section (e.g. Wonders, 1979, and this paper), but was assigned to the Late Cenomanian by Fazzini & Mantovani (1965) at Monte Subasio near Perugia. A similar black mudstone bed has been found in the southern Limestone Alps on the Trento Swell (Castellarin, 1972), in several pelagic sections such as at Cismon (Channell *et al.*, 1979) near Covelo and Aldeno (Fuganti, 1966; Bosellini *et al.*, 1978) and in the Lombardy Basin (Venzo, 1954). These occurrences are also at the Cenomanian/Turonian boundary. Thus it appears that the event which led to the development of the horizon occurred over a large portion of Northern Italy, at least in sequences which are thought to have been deposited in relatively shallow water on pelagic plateaus (i.e. perhaps less than 1000 m). Jenkyns (1980) has reviewed these and other occurrences.

The Bonarelli horizon may mark an important oceanographic event which occurred during Cenomanian/Turonian time and which influenced much of the Tethys and the North and South Atlantic. A widespread Cenomanian/Turonian "Oceanic Anoxic Event" has been recognized by Schlanger & Jenkyns (1976) who reviewed the literature and noted that similar horizons of sediment rich in organic carbon, which indicate anoxic or dysaerobic conditions on the sea floor, occur in pelagic deposits of the Subbetic belt of Spain, the chalks of England and even in areas of northern South America, the Caribbean and the mid-continent Cretaceous seaway of North America. Scholle & Arthur (1980) have shown that a major carbon isotope event, a shift to abnormally "heavy" $\delta^{13}C$ values, occurs in pelagic chalk and limestone near the Cenomanian/Turonian boundary in a number of sections from Europe and Mexico. This $\delta^{13}C$ event may be a reflection of the rapid increase in storage of isotopically light organic carbon in marine sediment associated with the "Oceanic Anoxic Event", or it may partly reflect a period of enhanced vertical stability in the water column with low rates of upwelling. In the latter case, more positive "heavier" $\delta^{13}C$ values would have occurred in surface water because of depletion of ^{12}C by organisms in the absence of upwelling. It is not yet known whether this Cenomanian/Turonian event represents a brief period of increased productivity, enhanced conditions of organic matter preservation at the sea floor or both. Schlanger & Jenkyns (1976) suggest that the Cenomanian/Turonian anoxic event represents an expansion of the midwater oxygen-minimum zone during a highstand of sea level. This is thought to have been due partly to increased productivity in expanded shelf sea areas.

The lithology and chemistry of the Bonarelli horizon at Gubbio and in sections from the Lombardy Basin have been examined in this study in order to provide some limitations on hypotheses previously discussed.

6.2. Bonarelli horizon at Gubbio—lithology

The thickness and lithology of the Bonarelli horizon (Figures 18 and 19) have been logged at several localities at and near Gubbio. The Bonarelli layer is considered in this study to extend from the dark gray-black chert bed above gray limestone at the base of the bed through the gray-green laminated and streaked chert layer at the reoccurrence of gray limestone beds above. The thickness of the Bonarelli horizon varies from about 65 cm at a locality 5 km from Gubbio to about 100 cm at Gubbio. The sequence of lithologies within the horizon is similar at all localities (Figure 18). At Gubbio, the Bonarelli horizon (Figure 19) occurs at 111 m in the measured section within a sequence of light-gray to creme-colored limestone beds, which are punctuated by dark-gray to black chert beds and nodules less than 6 cm thick and

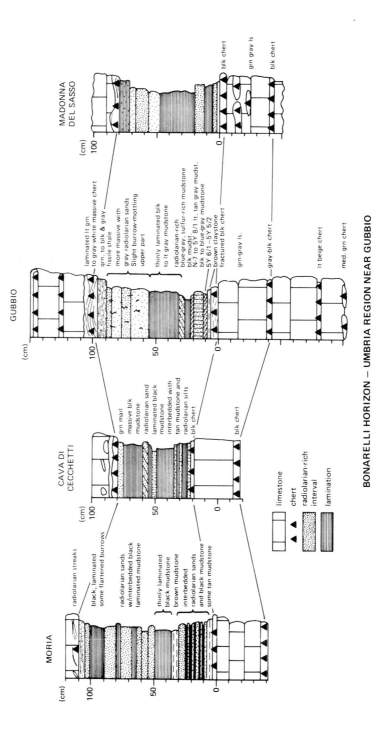

Figure 18.   Lithology of the lower Turonian "Bonarelli horizon" or Scisti a Ittiolitici at four localities in the Umbria region near Gubbio. Localities shown in Figure 1. Note radiolarian-rich intervals (silts). Sediment is also rich in amorphous organic matter (up to 22%) and phosphatic debris. Many intervals contain radiolarian silts/sand and mudstone interbedded on a centimeter scale.

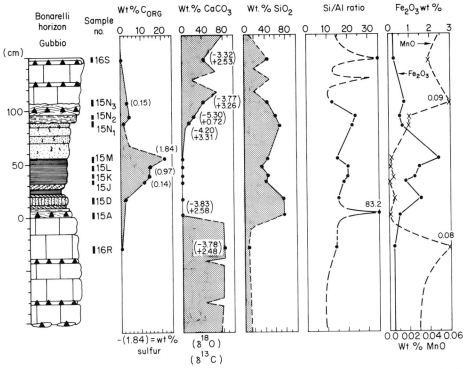

Figure 19.   Lithology and chemistry of samples from the "Bonarelli horizon" (111 to 112 m in the Gubbio section). (Data from Arthur, 1979.) Iron content is expressed as Fe$_2$O$_3$ although it is probably present mostly in a reduced state.

contain several thin beds of fissile black marlstone. Ten meters above the top of the Bonarelli horizon, the limestones and cherts become light-red to salmon-pink and variegated yellow, gray, light-link and creme, reflecting a change to a higher oxidation state and a resulting decrease in organic carbon and pyrite content. Within the Bonarelli layer, carbonate contents drop to nearly zero.

The limestone just below the Bonarelli Horizon is a homogeneous (non-laminated) light creme-colored, partly silicified foraminiferal micrite. Planktonic foraminifers are abundant and consist of both keeled and "large globigerinid" (*Whiteinella*) types. A few calcispheres and very rare calcareous benthic foraminifers are present. Fragments of planktonic foraminifers are common, and foraminifer tests are filled with micritic calcite. Radiolarians occur at the base of the Bonarelli micrite-cherty limestone. The gray-black chert within this bed is fractured, and fractures are filled with calcite. The limestone is micritic with no planktonic or benthic foraminifers visible. However, poorly-preserved radiolaria are abundant in parts of the limestone, especially within a centimeter of the chert layer. The radiolaria are filled with chalcedony in the limestone portion, but, within the silicified zone near the chert, radiolarian molds are often filled with calcite. Ghosts of radiolaria occur within the chert. Although no calcareous benthic foraminifers were noted in thin-section, an impoverished benthic infauna was present because rare flattened burrow mottles filled with black organic carbon-rich mud are preserved. The basal 30 cm of the Bonarelli horizon consists primarily of light-colored (ranging from tan to steel-gray) clayey, somewhat friable, muddy radiolarian sandy silt layers with relatively small amounts of carbonate and dark organic matter in the matrix. Individual radiolarian sand/silt layers show no signs of textural grading at Gubbio or at any of

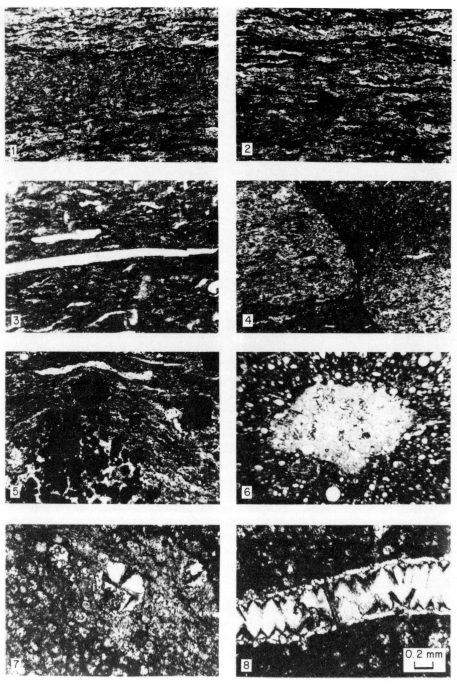

Figure 20. Photomicrographs of sediment from the "Bonarelli horizon". All at the same scale. (1 and 2)
Laminated, brown-black mudstone rich in organic carbon. Dark material is wispy amorphous
organic matter; light areas are laminae of flattened radiolaria now replaced by microcrystalline
chalcedony. Two scales of lamination are apparent in (1). Middle fine-grained zone is pseudoisotropic silica-phosphate mixture with less organic carbon than laminae above one below. (111 m, E.
Turonian, plane light.) (3 and 4) Laminated mudstone similar to (1) and (2) above. (3) shows example
of phosphatic debris abundant in portions of these brown-black mudstones. These are probably
cross-sections of fish scales. (4) illustrates a microfaulted laminated layer. Microfaults are commonly
visible in thin section and appear to have concentrations of oily organic matter along them (dark areas
in photo). (111 m, E. Turonian, plane light.) (5) Laminated brown-black mudstone, rich in organic
carbon with abundant early diagenetic pyrite which formed in siliceous interval of sediment
(originally laminae of radiolarian silt, now light-colored recrystallized areas within pyrite concentration). Note differential compaction and bending of mudstone laminae around pyrite crystals.
(111 m, E. Turonian, plane light.) (6) Calcite-cemented radiolarian siltstone in upper part of
Bonarelli layer. (111 m, E. Turonian, polarized light.) (7 and 8) Calcareous radiolarian siltstone.
Radiolarian walls replaced and tests filled with fibrous, radial chalcedony. (8) shows chalcedony-filled
fracture. (111 m, E. Turonian, polarized light.)

the other localities studied. The radiolaria are poorly preserved, having been replaced by fans of chalcedony (Figure 20); the tests range in diameter from 0.07 to 0.25 mm. The radiolarian-rich layers are generally not laminated and may have rare organic-rich burrow mottles within them. Some thin, laminated organic carbon-rich mudstone beds are interbedded with the 3–8 cm thick individual radiolarian layers (Figure 20).

The middle 30–35 cm of the Bonarelli horizon consists almost entirely of completely unbioturbated, very thinly-laminated black to dark olive-brown-black mudstone. Organic carbon values range from 11.8 to 21.7% in this interval. This compares well with measurements from the other localities shown in Figure 18 which gives values from 1.6 to 21.8% organic carbon. The organic matter occurs mainly as compressed amorphous wisps and streaks of amber to brown material disseminated in the matrix or as part of thin laminae couplets, alternating with thin laminae of siliceous and phosphatic material (Figure 20). The laminae range in thickness from 0.07–0.1 mm to 0.7–2.4 mm. Contacts between laminae are gradational to sharp. Phosphatic debris is abundant, comprising up to 30% of some sections. This occurs mainly as fish scales and vertebrae which commonly are found along partings in the mudstone in the field and in massive laminae of unstructured pseudoisotropic material. Phosphatic debris noted in this section is structured amber to white "collophane" with cross-sections up to 1.5 mm across. The muds have been compacted at least 40% judging by the bending of laminae around resistant particles such as phosphatic debris and the common pyrite blebs, cubes, and pyritohedra (clots up to 0.4 mm diameter). The siliceous material is largely unstructured, and the original siliceous organisms cannot be identified. Some round to ellipsoid cross-sections were seen in thin-section; these could be ghosts of radiolaria. Much of the siliceous material probably consists of crushed and altered radiolarian tests.

Microscopic examination of organic matter in the laminated black mudstones shows it to be entirely amorphous. Only a few very well-preserved pollen spores and no dinoflagellate cysts were found. The organic matter is presumably of marine phtoplanktic origin, as suggested by optical studies and pyrolysis results (Figure 16) which show high hydrogen and low oxygen indices for most samples.

These sections of several samples (Figure 20) show a fine vertical network of fractures or microfaults cutting the sediments along which are amber bituminous concentrations resembling oil stains. It is possible that hydrocarbons were generated *in situ* in the bed and began to migrate vertically. However, no reservoir rocks were present for accumulation, and the hydrocarbon substance remained with the rock. Barnaba (1958) reports having found oily material in samples he examined. Apparently, the organic carbon-rich material of the Bonarelli level was mined in small quantities and used by early inhabitants of Gubbio for heating fuel (Ettore Sannipoli, pers. commun.); it burns slowly with an oily smoke.

The organic carbon-rich mudstone interval is overlain by medium to light gray and then light creme-colored radiolarian mudstone and radiolarian sandy silt layers. This interval is about 40 cm thick. The organic carbon content is relatively low (from 1 to 4%) compared to the laminated mudstone interval (Figure 20). The carbonate content increases gradually toward the top of the bed. Most of the carbonate occurs as a patchy micritic matrix and as some sparry calcite cement and vein fillings. Radiolarians are poorly preserved, and the tests are filled with fibrous, radiating masses of chalcedony. The test walls are sometimes replaced by micritic calcite matrix. Pyrite is abundant and often occurs along walls of replaced radiolarian tests. The mixture of radiolarian sizes, ranging from medium silt to fine sand, shows no textural grading in the sections studied. Large (up to 1 cm diameter) organic carbon-filled furrows are present but rare in the radiolarian layers. Minor amounts of barite, clinoptilolite and dolomite were detected in x-ray diffraction traces of silica-rich samples.

The Bonarelli horizon is capped by a bed of creme-colored silicified limestone and chert. The chert is faintly laminated with streaks and wisps of radiolaria. Flattened burrow mottles are common. No planktonic or benthic foraminifers were noted in a thin section of the radiolarian micrite. Two generations of fractures are present: the first is filled with chalcedony, the later cross-cutting one by calcite.

6.3.   Rate of deposition and duration of anoxic event

The Bonarelli horizon represents a significant period of enhanced preservation of organic matter in the pelagic sediment sequence at Gubbio. Knowledge of the rate of deposition and duration of the event is important to understand the origin of this unusual sedimentary horizon. Three possible approaches are here discussed as indices to rate of deposition: a possible varving, average rates, and the use of aluminum and of titanium contents.

The middle 30–35 cm of the 100 cm thick layer are very finely laminated and rich in organic carbon (up to 22%). The average organic carbon-silica couplet or laminae is about 0.01 cm thick. The laminae are difficult to count because they are often very thin and lenticular, but if the laminae represent annual varves, the 35 cm interval represents a minimum of 3500 years. This would indicate that the sediment of the Bonarelli horizon was deposited at a rate of more than twenty times that of the limestone and chert above and below. In this case, the layer could be due to dramatically-increased surface productivity of organic matter and biogenic silica. However, this hypothesis rests on the assumption that the varves are annual, which may not be correct.

At an average, calculated sedimentation rate of 5.3 m per m.y. (0.53 cm per 10^3 years) for the Turonian (Arthur, 1979b), the Bonarelli layer would represent about 0.21 m.y. However, this may be unreasonably low because the supply of carbonate was apparently cut off during deposition of the 110 cm layer. If one assumes that there was no increase in rate of accumulation or composition of the terrigenous component, then aluminum and/or titanium contents can be used as an indicator of the overall sedimentation rate. The average Al_2O_3 contents for limestones below (85.5–110.0 m) and above (112.0–143.0 m) the Bonarelli horizon are 0.82 and 0.76% respectively (Arthur, 1979b). These values are remarkably constant through the sequence which was deposited at rates (compacted) of 4.8–5.3 m per m.y. (0.48–0.53 cm per 10^3 years). The calculated accumulation rate of Al_2O_3 is between 1.04 and 1.07 g cm^{-2} per 10^3 years. The average Al_2O_3 content of sediment in the Bonarelli horizon is 2.42%. If the accumulation rate of Al_2O_3 is assumed to be equivalent to that in the limestone sequence above and below, then this gives a total sedimentation rate of 1.6 m per m.y. (0.16 cm per 10^3 years) for the Bonarelli bed. This rate is one-third that of the limestone sequence and suggests that the entire 100 cm unit accumulated in about 0.7 m.y. and that the major period of anoxia and organic carbon preservation (around 35 cm of sediment) lasted about 0.25 m.y.

A similar calculation using titanium contents gives an intermediate figure of about 0.35 m.y. for the time of deposition of the entire Bonarelli horizon. The most realistic estimate probably lies between 0.35 and 0.7 m.y.

6.4.   Paleoenvironmental implications

The Bonarelli bed is unusually rich in silica, organic carbon and phosphatic debris and impoverished in calcium carbonate relative to the typical strata of Cenomanian–Turonian age at Gubbio. Organic carbon values are up to 100 times higher than in adjacent limestones (Figure 9) and phosphate is greatly increased. The SiO_2 content is very high partly because there is no dilution by carbonate, but when normalized to aluminium content (Figure 19), the silica values are still more than two times the normal value. This "excess" silica is derived from biogenic silica and may indicate a relative increase in surface productivity and/or preservation of radiolarian tests in sediment. Berger (1976) has summarized evidence that suggests that preservation of siliceous tests in sediment is primarily dependent on the rate of production in surface waters.

The organic matter is amorphous and probably of marine origin as previously mentioned. The greatly enhanced preservation of organic matter, the fine lamination in the sediment, the lack of burrowing, and the apparent absence of benthic organisms suggest that anoxic conditions existed at or above the sea floor. Conditions certainly were anoxic within the sediment. Pyrite is very common, and contents of up to 1.84% sulfur were measured in organic carbon-rich samples (Figure 19). This value is higher than average values of Holocene marine sediment rich in organic carbon (e.g. Kaplan et al., 1963; Goldhaber and Kaplan, 1974). The maximum sulfur measured in anoxic Gulf of California sediment is about 2% (Berner, 1964). Samples were chosen as carefully as possible to eliminate the effects of outcrop weathering on chemical analysis. Even so, in the Bonarelli horizon, it is very difficult to sample very fresh material because the layer in all exposures appears to have been a conduit for post-uplift groundwater circulation. A stain of leached iron oxide and free sulfur is present on outcrop surfaces on and below the bed. Therefore, original organic carbon and sulfur values may have been even higher than measured. Some carbonate within the upper part of the bed may have been dissolved and reprecipitated by groundwater based on low Sr^{2+} values and more negative $\delta^{18}O$ and $\delta^{13}C$ values (Arthur, 1979b; Figure 19).

Iron and manganese contents also indicate the original oxidation state of the layer. Manganese is nearly absent within the intervals high in organic carbon in the Bonarelli horizon, but MnO is present in concentrations of as much as 0.08 and 0.09% below and above the layer (Figure 19). These values are three times the average MnO content within adjacent limestone beds and may represent a fossilized "oxidation front" where manganese was dissolved from within the Bonarelli bed and migrated into the surrounding more highly oxidized sediment where it was precipitated as an oxide. Iron values (as Fe_2O_3) in the Bonarelli horizon are up to ten times the average background level in adjacent reduced sedimentation rates within the interval. Much of this iron is probably present as sulfide minerals. Values of trace elements such as Co, V, Cu, Cr and especially Zn are also higher in the Bonarelli layer than in adjacent limestone beds, typically higher than would be expected for the relatively reduced total accumulation rates. These elements may be associated with organic matter or sulfides and further indicate anoxic conditions (Calvert, 1976; Price, 1976, for reviews).

The data discussed above suggest that sedimentation of much of the Bonarelli horizon took place under anoxic conditions on the sea floor, and that the "anoxic event" lasted at least 0.25 m.y. This seems an unusually long period of anoxia for a pelagic sea floor. This early Turonian "anoxic event" took place just prior to a relatively long period of apparent higher oxygen supply to the sea floor and sediments (late Turonian–Paleocene). These "events" occur within a sedimentary sequence (gray limestone, somewhat laminated and organic carbon-rich) which suggests a generally low level of oxygenation of bottom water. Evidence from the Gubbio section alone does not constrain any hypotheses of the origin of anoxic bottom waters—whether they represent an expanded or intensified oxygen-minimum layer (Schlanger & Jenkyns, 1976; Fischer & Arthur, 1977; Thiede & van Andel, 1977) impinging on a slope and deep outer pelagic shelf, or the stagnation of an entire basin due to high salinity contrasts between surface and deep water masses (e.g. Ryan & Cita, 1977). The characteristics of the Bonarelli horizon sediment are similar to those from slopes under oxygen-minimum zones in high productivity regions (e.g. Calvert, 1964; Berger & Soutar, 1970; von Stackelberg, 1971; Burnett, 1977). The high silica and phosphate contents and low carbonate values are typical of the sediments of the Peru–Chile margin and the Gulf of California. However, differences are apparent between the modern examples. Berger & Soutar (1970) show that carbonate preservation is actually enhanced in sediment of some anoxic basins (e.g. Santa Barbara) due to the alkalinity increase resulting from sulfate reduction. In the case of the Bonarelli horizon, it is not clear whether carbonate was dissolved in caustic bottom waters or was not produced in surface waters. However, the abundance of *Whiteinella* ("large globigerinids") and the paucity of keeled planktonic foraminifers at or near the Cenomanian–Turonian boundary at Gubbio (in fact such relations are widespread in Tethys and the Western Interior U.S.) suggest an unstable, possibly upwelling environment. The excess silica and radiolarian-rich beds suggest upwelling and high productivity during deposition of the Bonarelli layer, especially near the beginning and end of the event which produced it. Although increased production and preservation of biogenic silica is typical of upwelling zones (Berger, 1975; Diester Haass & Schrader, 1979), von Stackelberg (1972) found little or no biogenic silica in sediment from the slope off Pakistan and attributed that absence to dissolution in acidic (*sic*) waters with low oxygen content. It is not clear in that case what mechanism von Stackelberg intended since opal is more soluble under alkaline conditions.

The widespread occurrence of similar layers rich in organic carbon near the Cenomanian/Turonian boundary in other sections of ancient Tethys (Schlanger & Jenkyns, 1976) indicates that the event which led to the "Livello Bonarelli" was not local. Most of these examples come from DSDP sites in relatively shallow water (i.e. less than 2500 m) or from sequences exposed on land which show evidence of having been deposited on slopes or outer pelagic shelves (Schlanger & Jenkyns, 1976). A hiatus or slowly-deposited oxidized sediment characterized the Cenomanian–Turonian boundary interval in most deeper water oceanic sequences cored at DSDP sites (Arthur & Schlanger, 1979; Arthur & Natland, 1979) except possibly along the continental margin of North-West Africa (Wiedmann *et al.*, 1978). It appears from this evidence that the Cenomanian/Turonian "Oceanic Anoxic Event" represents a brief period of expansion and intensification of a mid-water oxygen minimum. This event was apparently widespread and most intense in the circum-equatorial Tethys and Pacific (Schlanger & Jenkyns, 1976), but it affected waters at higher paleolatitudes as well, including parts of the seas of England (Jeffries, 1962, 1963) and the Western Interior Seaway of North America (Frush & Eicher, 1975), possibly carried into these shelf seas by a major marine transgression (e.g. Schlanger & Jenkyns, 1976).

In most of the Southern Alps, white-colored Upper Albian Scaglia Bianca or equivalent formations are capped by reddish Cenomanian pelagic or hemipelagic strata. These, in turn, are capped by black shaly or marly limestone layers, rich in radiolarians, ranging in thickness from 1 m up to 7 m. In some sequences, they are very bituminous and somewhat rich in plant debris.

The predominantly red Cenomanian strata, locally called Scaglia Rossa (Breggia and Lune sections), the Sommaschio Formation, or the Scaglia Variegata, again differ from one locality to another in carbonate content: the Sommaschio Formation, at the Torre de' Busi locality, originally in the deepest part of the Lombardy Basin, is maroon in color and is devoid of autochthonous carbonates; pelagic carbonate occurs only in clearly resedimented layers. A few dark organic carbon-rich layers of 10 cm or more thickness are scattered throughout the formation; some of them are clearly older than the Bonarelli Layer, but the highest Corg-rich layers possibly are equivalent to the Bonarelli. The Cenomanian strata appear to be relatively rich in carbonates in all the other sections, although the carbonates have been more diluted by clay than in the underlying Scaglia Bianca equivalents. Black-shale layers in these sections appear to be confined to the Cenomanian–Turonian boundary. In slope settings, the black bed can be a few meters thick, (e.g. at Gavarno, Pisgia; Gelati & Cascone, 1981), while in pure pelagic sequences, such as in the Trento area (Bosellini *et al.*, 1978), the Euganei Hills (Sorbini, 1976) or in the Venetian Alps, the black shales are about 1 m thick, similar to the "Bonarelli". Differences in thickness are probably related to variations in the rate of sedimentation which varies from place to place in this interval.

Large slump units, breccias and conglomerates are recorded in the vicinity of the black shale layer; these may be a consequence of tectonic movements related to the pre-Gosau orogenic phase which affected the Southern Alps. After the Cenomanian–Turonian black shale episode, flysch sedimentation was widespread in the Lombardy Basin; these sediments were deposited mainly in poorly-oxygenated basinal environments (e.g. at Breggia, Pontida and Torrente Sommaschio). In the more elevated areas, such as at Tignale (Lune) and M. Baldo, the pelagic sediments are devoid of clastic input and resemble those from the Umbria region containing abundant chert, as in Gubbio, to the end of the Santonian (Tignale section, Cita, 1948; Castellarin, 1972). In pelagic sequences, the Cenomanian–Turonian black shale layer is rich in radiolarians and poor in carbonate and has a stratigraphy similar to that of the Bonarelli.

At the Cenomanian–Turonian boundary, the amount of terrestrial organic material in the total organic fraction increases considerably in the Lombardian sections along with the strong increase in terrigenous sediment input from larger areas under erosion; as mentioned above, by Bonarelli time, the Lombardy Basin was already under a flysch sedimentary regime (Venzo, 1954; Castellarin, 1977; Aubouin *et al.*, 1970; Gelati & Cascone, 1981). Nevertheless, the highest Corg content in the Southern Alpine section coincides with an increase of amorphous organic matter similar to that in the Bonarelli Horizon. The amorphous material is much more diluted by terrestrial material, but the marine input remains visible. The Corg content for measured samples are plotted in Table 1.

We still do not understand the causes of this expansion of the oxygen-minimum zone nor its apparent correlation with the eustatic sea level rise at the Cenomanian–Turonian boundary. One possibility is that this event represents a relatively rapid change in the density structure of the oceans accompanied by sudden overturn of intermediate to deep waters. The upwelling of these deep waters, which were probably nutrient rich, stimulated productivity and encouraged luxury extraction of silica by radiolaria. The deep waters could have been part of an old oxygen-depleted water mass which had, by virtue of a relatively long residence time, accumulated nutrients, CO_2 and dissolved silica. Oxygenation of deep waters in Albian–Cenomanian oceans was relatively poor, possibly because of stable stratification due to high salinity contrasts between the deep waters and surface waters or to the decreased solubility of oxygen in warm, saline deep-water masses. Following the relatively brief Cenomanian–Turonian event, the deep and intermediate water masses of the major ocean basins appear to have been better oxygenated, possibly due to more rapid turnover of deep water (Fischer & Arthur, 1977) or to a change in source of deep water. Although the most intense event, evidenced by the Bonarelli horizon, may have spanned no more than 0.7 m.y. at Gubbio, the overall period of intensified upwelling and organic carbon preservation may have been about 2 m.y. based on interpretation of $\delta^{13}C$ curves from pelagic sediment sequences, including the Gubbio sequence (Scholle & Arthur, 1980). Thin, organic carbon-rich partings occurring in the Upper Cenomanian below the Bonarelli layer at Gubbio also attest to this.

7. Conclusions

The widespread Cretaceous "Oceanic Anoxic Events" proposed, for example, by Schlanger & Jenkyns (1976) have their expression in pelagic sequences of the Cretaceous Tethyan seaway. Thin organic carbon-rich marlstones or shales are interbedded with pelagic limestones as old as Hauterivian in the Umbrian sequences, and the Maiolica Limestone and equivalents of Barremian through Early Aptian age particularly are characterized by interbedded thin marlstones containing up to 4% organic carbon and bioturbated, often cherty, gray, micritic limestones. Carbonate accumulation rates in most sections appear to have declined markedly in the Latest Aptian through Middle Albian time, while terrigenous flux increased slightly. This led to deposition of relatively thin sequences (<50 m) of marlstones and marly limestones in the Umbrian region and thicker sequences (80–220 m) in the Southern Alps. These sequences are characterized by rhythmicity of interbedded dark, relatively organic carbon-rich and often laminated black claystones or marlstones and bioturbated, light-colored marlstone and limestone beds. The rhythmicity is of the order of 20 000 to 50 000 years. Organic carbon content ranges from 0–15% in the Aptian–Albian sequences, but averages about 1.0%. Poorly-preserved radiolaria, frequently pyritized, are often concentrated in discrete layers. The lithologic, paleontologic and chemical relationships suggest periodic upwelling as the origin of the cyclicity. Deep water structure was probably stable and oxygen contents relatively low. Our data do not distinguish between an expanded oxygen-minimum model and an entirely anoxic or dysaerobic deep-basin model. However, organic carbon-rich strata are found at all inferred paleodepths from shallow plateau to slope and on oceanic crust. Therefore, it is likely that this part of Tethys was poorly oxygenated throughout the water column. Productivity was not necessarily high overall.

Upper Albian pelagic sedimentary rocks evidence an increase in rate of carbonate accumulation, carbonate content, and in oxygenation at the sea floor. The Early to Middle Cenomanian in most settings was characterized by deposition of light-colored pelagic limestones verging on shades of pink and red. Relatively little organic carbon was preserved in sediments of this age.

However, in Late Cenomanian time, productivity apparently increased, accompanied by a relative decrease in oxygenation of deep-water masses as evidenced by a few thin organic carbon-rich shales interbedded with gray, sometimes laminated pelagic limestones. A major upwelling event occurred at the Cenomanian–Turonian boundary resulting in the deposition of a very organic carbon- and radiolarian-rich bed (up to 23% Corg) between 50 and 100 cm thick in the Umbrian region. The organic carbon content is somewhat diluted by terrigenous material in Southern Alps sections. The upwelling and high productivity event is estimated to have lasted approximately 0.25 m.y. Bottom waters apparently became well oxygenated by the Late Turonian as evidenced by red-colored limestones (Corg $<0.01\%$) in most sections, except in the deepest troughs of the Lombardy Basin.

Organic matter in the Cretaceous organic carbon-rich intervals studied here is primarily of marine derivation. Pyrolysis data suggests that layers containing less than about 1% Corg are dominated by terrigenous organic carbon and/or oxidized marine amorphous organic matter. Pyrolysis results also suggest that much of the organic matter in the outcrop sections is relatively immature in agreement with vitrinite reflectance determinations (e.g. 0.4–0.6% R). However, in some deeply-buried sequences, the Lower Cretaceous section is a potential hydrocarbon source rock.

Acknowledgments

We are indebted to numerous colleagues and students for help in the field, for discussion of problems related to the origins of pelagic sequences rich in organic carbon, and for some of the analyses reported here. For constant and helpful discussion and criticism, we particularly thank M. B. Cita, G. E. Claypool, W. E. Dean, A. G. Fischer, H. C. Jenkyns, S. O. Schlanger and P. A. Scholle. W. E. Dean, A. G. Fischer and H. C. Jenkyns provided critical reviews of an earlier manuscript. We thank them for their comments which improve the manuscript and expression of ideas. Some chemical analyses were provided by Tiberto (Instituto di Mineralogia e Petrografia, University of Milan) and E. Bonatti (RSMAS, University of Miami); pyrolysis analyses of our samples were supplied by G. Deroo (Institute France du Petrole) and G. E. Claypool (USGS, Denver); some organic carbon analyses were performed by Drs Grignani and Biffi (AGIP); vitrinite reflectance data was courtesy of N. Bostick (USGS, Denver); palynofacies analysis was by R. Ferraris and G. Piccaia (University of Milan). We also thank L. Paggi, E. Mainardi, R. Gelati, and E. Sannipoli for paleontological data and for help in the field.

This is an official contribution of IGCP Project No. 58—"Mid-Cretaceous Events". We thank M. B. Cita and S. O. Schlanger for inviting our participation in the symposium. Our research was supported partly by grants from the Consiglio Nazionale della Ricerche, Comitato 05 to Gruppo Informale "Paleopelagos" and NSF Grant DES 74-22214 to A. G. Fischer.

References

Abbate, E., Bortolotti, V., Passerini, P. & Sagri, M. 1970. Introduction to the geology of the Apennines. *Sedimentary Geology* **4**, 207–249.

Abbate, E., Bortolotti, V. & Passerini, P. 1972. Studies on mafic and ultramafic rocks: 2—Paleogeographic and tectonic considerations on the ultramafic belts in the Mediterranean area. *Boll. Soc. Geol. Ital.* **91**, 239–282.

Abbate, E. & Sagri, M. 1970. The eugeosynclinal sequences. Development of the Northern Apennines Geosyncline. *Sedimentary Geology* **4**, 251–340.

Andri, E. & Fanucci, F. 1975. La resedimentazione dei Calcari Calpionella ligure. *Boll. Soc. Geol. Ital.* **94**, 915–925.

Arthur, M. A. 1979a. North Atlantic Cretaceous black shales: the record at Site 398 and a brief comparison with other occurrences. *Initial Reports of the Deep Sea Drilling Project* **47**, (2), 719–751.

Arthur, M. A. 1979b. Sedimentologic and geochemical studies of Cretaceous and Paleogene sedimentary rocks and some global paleoceanographic trends and events. Part I: the Gubbio sequence. *Ph.D. Dissertation*, Princeton University, 171pp.

Arthur, M. A. & Jenkyns, H. C. 1981. Phosphorites and paleoceanography. (Ed.W. H. Berger). Ocean geochemical cycles, *Proceedings 26th International Geological Congress, Oceanologica Acta*, Spec. Issue, pp. 83–96.

Arthur, M. A. & Natland, J. H. 1979. Carbonaceous sediments in the North and South Atlantic: the role of salinity in stable stratification of early Cretaceous basins. (Eds M. Talwani, W. W. Hay & W. B. F. Ryan). *Results of Deep Drilling in the Atlantic Ocean, Proceedings of the Second Maurice Ewing Symposium (1978)*, Vol. 3, pp. 375–401.

Arthur, M. A. & Schlanger, S. O. 1979. Cretaceous "oceanic anoxic events" as causal factors in development of reef-reservoired giant oil fields. *American Association of Petroleum Geologists Bulletin* **63**, 870–885.

Aubouin, J. 1963. Essai sur la paleogeographie post-triasique et l'évolution secondaire et tertiare du versant sud des Alpes Orientales (Alpes Méridionales; Lombardie et Venétie, Italie; Slovenie Occidentale, Yugoslavie). *Bull. Soc. Géol. France, Ser. 7* **5**, 730–766.

Aubouin, J., Sigal, J., Berland, J. P., Blanchet, R., Bonneau, M., Cadet, J. P., Guilliot, P. L., Lacour, A., Piat, B. & Vicente, J. C. 1972. Sur le bassin de Flysch: stratigraphie et paleogeographie des Flyschs cretaces de la Lombardie (versant sud des Alpes orientales, Italie). *Bull. Soc. Geol. France, Ser. 7* **12**, 612–658.

Barnaba, P. F. 1958. Geologia dei Monti di Gubbio. *Boll. Soc. Geol. Ital.* **77**, 39–70.

Berger, W. H. 1975. Deep-sea carbonates: dissolution profiles from foraminiferal preservation. *Cushman Found. Foram. Res., Special Publication* **13**, 82–86.

Berger, W. H. 1976. Biogenous deep-sea sediments: production, preservation and interpretation. *Treatise on Chemical Oceanography*. (Eds J. P. Riley & R. Chester. Vol. 5. London and New York: Academic Press, pp. 265–388.

Berger, W. H. 1979. Impact of deep sea drilling on paleoceanography: in *Results of Deep Drilling in the Atlantic Ocean: Second Maurice Ewing Symposium*. (Eds M. Talwani, W. W. Hay & W. B. F. Ryan. Vol. 3, American Geophysical Union, p. 297–314.

Berger, W. H. & Johnson, T. C. 1976. Deep-sea carbonates: Dissolution and mass wasting on Ontong-Java Plateau. *Science* **192**, 785–787.

Berger, W. H. & Soutar, A. 1970. Preservation of plankton shells in an anaerobic basin off California. *Geological Society of America Bulletin* **81**, 275–282.

Berner, R. A. 1964. Distribution and diagenesis of sulfur in some sediments from the Gulf of California. *Marine Geology* **1**, 117–140.

Bernoulli, D. 1972. North Atlantic and Mediterranean Mesozoic facies: a comparison. *Initial Reports Deep Sea Drilling Project* **11**, 801–871.

Bernoulli, D. & Jenkyns, H. C. 1974. Alpine, Mediterranean, and Central Atlantic Mesozoic facies in relation to the early evolution of the Tethys: *Society of Econ. Paleontological and Mineral. Special Publication* **19**, 129–160.

Bernoulli, D., Kälin, O. & Patacca, E. 1979. A sunken continental margin of the Mesozoic Tethys: the northern and central Apennines. *Symposium on Sedimentation Jurassique W. Européen, A. S. F. Publication Speciale* **1**, 197–210.

Bitterli, P. 1963. Aspects of the genesis of bituminous rock sequences. *Geologie en Mijnbouw* **42**, 183–201.

Bitterli, P. 1965. Bituminous intercalations in the Cretaceous of the Breggia River, S. Switzerland. *Bull. Ver. Schweiz. Petrol.—Geol. u.—Ing.* **31**, 179–185.

Boccaletti, M. and Sagri, M. 1966. Brecce e lacune al passaggio Maiolica-Gruppo degli scisti Policromi. *Val di Lima Mem. Soc. Geol. It.* **5** (1), 19–66.

Bortolotti, V., Passerini, P., Sagri, M. & Sestini, G. 1970. The miogeosynclinal sequences. *Sedimentary Geology* **4**, 341–444.

Bosellini, A. 1973. Modello geodinamico e paleotettonico delle Alpi meridionali durante il Giurassico-Cretacico, Sue possibili applicazioni agli Appennini. *Acc. Naz. Lincei, Quadeno* **183**, 165–213.

Bosellini, A., Broglio Loriga, C. & Busetto, C. 1978. I bacini cretacei del Trentino. *Riv. Ital. Paleont. Strat.* **8414**, 897–946.

Burnett, W. C. 1977. Geochemistry and origin of phosphorite deposits from off Peru and Chile. *Geological Society of America Bulletin* **88**, 813–823.

Burnett, W. C., Veeh, H. H. & Soutara, A. 1980. U-series, oceanographic, and sedimentary evidence in support of recent formation of phosphate nodules off Peru. *Marine Phosphorites, Soc. Econ. Paleontol. Spec. Publ.* **29**. (Ed. Y. Bentor), 61–71.

Calvert, S. E. 1964. Factors affecting the distribution of laminated diatomaceous sediments in Gulf of California. *American Association of Petroleum Geologists Mem.* **3**, 311–330.

Calvert, S. E. 1976. The mineralogy and geochemistry of nearshore sediments. *Chemical Oceanography*. (Eds J. P. Riley & R. Chester). 2nd edition, Vol. 6. New York: Academic Press, 187–280.

Canuti, P. & Marcucci, M. 1967. Lacune della serie Toscana. IV—Osservazioni sui rapporti stratifrafici tra Maiolica e Scisti Policromi nella Toscana Centro-meridionale (Area di Rapolano). *Boll. Soc. Geol. It.* **86** (4), 809–818.

Castellarin, A. 1972. Evoluzione paleotettonica sinsedimentaria al limite tra "Piattaforma Veneta" e "Bacino Lombardo" a Nord di Riva del Garda. *Giorn. Geol.* **38**, 11–212.

Castellarin, A. 1977. Ipotesi paleogeografica sul Bacino del Flysch Sudalpino Cretacico. *Boll. Soc. Geol. It.* **95**, 501–511.

Centamore, E., Chiocchini, M., De Iania, G., Micarelli, A. & Pieruccini, U. 1971. Contriduto alla conoscenza dal Giurassico dell' Appennino Umbro-marchigiano. *Studi Geol. Camerti* **1**, 7–90.

Channell, J. E. T. & Horvath, F. 1976. The African/Adriatic promontory as a paleogeographical premise for Alpine orogeny and plate movements in the Carpatho–Balkan region. *Tectonophysics* **35**, 71–101.

Channell, J. E. T., Lowrie, W. & Medizza, 1979. Middle and Early Cretaceous magnetic stratigraphy from the Cizmon section, Northern Italy. *Earth and Planetary Science Letters* **42**, 153–166.

Cita, M. B. 1948. Ricerche stratigrafiche e micropaleontologiche and Cretacico e sull'Eocene di Tignale (Lago di Garda). *Rivista Ital. Paleont.* **54**, 54pp.

Cita-Sironi, M. B. 1964. Recerche micropaleontologiche e stratigrafiche suit sedimenti pelgaci del Giurassico superiore e del Cretacico inferiore della catena del Monte Baldo. *Riv. It. Paleont. Strat. Mem.* **10**, 1–160, 12 tav., 40 fig. Milano.

Colacicchi, R. & Praturlon, A. 1965. Stratigraphical and paleogeographical investigation on the Mesozoic shelf-edge facies in Eastern Marsica (Central Apennines, Italy). *Geol. Rom.* **4**, 89–118.

D'Argenio, B., De Castro, P., Emiliani, C., Simone, L. 1975. Bahamas Apennines identical mesozoic lithofacies. *American Association of Petroleum Geologists Bulletin* **59**, 524–530.

Dean, W. E., Gardner, J. V., Jansa, L. F., Cepek, P. & Seibold, E. 1978. Cyclic sedimentation along the continental margin of northwest Africa. *Initial Reports Deep Sea Drilling Project* **41**, 965–989.

de Graciansky, P. C., Auffret, G. A., Depeuble, P., Montadert, L. & Müller, C. 1979. Interpretation of depositional environments of the Aptian/Albian black shales on the north margin of the Bay of Biscay. *Initial Reports of the Deep Sea Drilling Project* **48**, 877–908.

Demaison, G. J. & Moore, G. T. 1980. Anoxic environments and oil source bed genesis. *American Association of Petroleum Geologists Bulletin* **64**, 1179–1209.

Dewey, J. F., Pitman, W. C., III, Ryan, W. B. F. & Bonnin, J. 1973. Plate tectonics and the evolution of the Alpine system. *Bulletin of the Geological Society of America* **84**, 3137–3180.

Diester Haass, L. & Schrader, H. J. 1979. Neogene coastal upwelling history off Northwest and Southwest Africa. *Marine Geology* **29**, 39–53.

Elter, P. 1972. La zona ofiolitifera del Bracio mel quadro dell'Appennino settentronale: Guida alle escursioni, 66, *Congresso. Soc. Geol. Ital.*, Pisa, 63pp.

Fazzini, P. & Mantovani, M. P. 1965. La Geologia del Gruppo di M. Subasio. *Boll. Soc. Geol. Ital.* **84**, 71–142.

Fischer, A. G. & Arthur, M. A. 1977. Secular variations in the Pelagic Realm. *Deep water carbonate environments, Soc. Econ. Paleont. and Mineral. Spec. Pub.* (Eds H. E. Cook and P. Enos) **25**, 19–50.

Frush, M. P. and Eicher, D. L. 1975. Cenomanian and Turonian foraminifera and paleoenvironments in the Big Bend region of Texas and Mexico. *The Cretaceous System in the Western Interior of North America: Special Paper Geological Association Canada*, **13**, 277–301.

Fuganti, A. 1961. Ricerche geologische & sedimentologiche sugli "scisti bituminosi uraniferi" de Mollaro (Val di Non-Trentino). *Studi Trenteni Sci. Nat.* **38** (1) 17–33.

Fuganti, A. 1966. Il tettonismo Cretacico e la deposizione degli "scisti neri uraniferi" del Trentino. *Atti del Symposium Internazionale sui giacimenti minerari delle Alpi* **2**, Trento-Mendola, 341–346.

Garrels, R. M. & MacKenzie, F. T. 1971. *Evolution of Sedimentary Rocks*. New York: W. W. Norton and Co., 395pp.

Gelati, R. & Cascone, A. 1981. Le successioni terrigene cretaciche delle Bergamasca. *Rendiconti Soc. Geol. Ital* **3** 39–40.

Ghelardoni, R. & Maioli, P. 1959. Stratigrafia e tettonica del M. Acuto—M. Filoncio (Umbria). *Boll. Serv. Geol. Italia* **80**, 215–222.

Goldhaber, M. B. & Kaplan, I. R. 1974. The sulfur cycle. *The Sea* (Ed. E. D. Goldberg) Vol. 5, 569–655.

Habib, D. 1979. Sedimentary origin of North Atlantic Cretaceous palynofacies. *Second Maurice Ewing Symposium, Results of Deep Drilling in the Atlantic* (Eds M. Talwani, W. W. Hay, & W. F. B. Ryan) Vol. 3, American Geophysical Union, 420–437.

Hesse, R. 1975. Turbiditic and non-turbiditic mudstone of Cretaceous flysch sections of the east Alps and other basins. *Sedimentology* **22**, 387–416.

Hsü, K. J. 1976. Paleoceanography of the Mesozoic Alpine Tethys. *Geol. Soc. Amer. Spec. Pap.* **170**, 44pp.

Jefferies, R. P. S. 1963. The stratigraphy of the *Actinocamax Plenus* subzone (Turonian) in the Anglo-Paris Basin. *Proceedings of Geological Association* **74**, 1–31.

Jenkyns, H. C. 1980. Cretaceous anoxic events: from continents to oceans. *Journal of Geological Society of London* **137**, 171–188.

Kaplan, I. R., Emery, K. O. & Rittenberg, S. C. 1963. The distribution and isotopic abundance of sulfur in recent marine sediments of Southern California. *Geochimica Cosmochimica Acta.* **27**, 297–331.

Lowrie, W., Alvarez, W., Premoli Silva, I. & Monechi, S. 1980. Lower Cretaceous magnetic stratigraphy in Umbrian pelagic carbonate rocks. *Geophysical Journal* **60**, 263.

McCave, I. N. 1979. Depositional features of organic-rich black and green mudstones at DSDP Sites 386 and 387, western North Atlantic. *Initial Reports of the Deep Sea Drilling Project* **43**, 411–416.

McIver, R. D. 1975. Hydrocarbon occurrences from JOIDES Deep Sea Drilling Project. *Ninth World Petroleum Congress, Proceedings*, Vol. 2, 269–280.

Merla, G. 1951. Geologia dell'Appennino Settentrionale. *Boll. Soc. Geol. It.* **70** (1), 45–382.

Müller, P. J. & Suess, E. 1979. Productivity, sedimentation rate, and sedimentary organic matter in the oceans—organic preservation. *Deep Sea Research* **26A**, 1347–1362.

Ogniben, L., Parotto, M. & Praturlon, A. (Eds) 1975. Structural model of Italy (maps and explanatory notes). *Quad. Ricerca Scientifica* **90**, 1–502.

Pasquaré, G. 1965. Il Giurassico superiore nelle Prealpi Lombarde. *Rivista Ital. Paleonto. Strat., Mem.* **11**, 1–228.

Praturlon, A. & Sirna, G. 1976. Ulteriori dati sul margine Cenomaniano della piattaforma carbonatirea Laziale-Abruzzese. *Geol. Rom.* **15**, 83–111.

Premoli Silva, I. 1977. Upper Cretaceous-Paleocene magnetic stratigraphy at Gubbio, Italy, II. Biostratigraphy. *Geological Society of America Bulletin* **88**, 371–374.

Premoli Silva, I. & Boersma, A. 1977. Cretaceous planktonic foraminifers— DSDP Leg 39 (South Atlantic). *Initial Reports of the Deep Sea Drilling Project* **39**, 615–641.

Premoli Silva, I. & Paggi, L. 1976. Cretaceous through Paleocene biostratigraphy of the pelagic sequence at Gubbio, Italy. *Mem. Soc. Geol. It.* **15**, 12–32, Pisa.

Premoli Silva, I. & Paggi, L. (in press). Mid-Cretaceous organic-rich sediments from the Italian pelagic sequences. *Rivista Ital. Paleont. Strat.*

Premoli Silva, I., Paggi, L. & Monechi, S. 1977. Cretaceous through Paleocene biostratigraphy of the pelagic sequence at Gubbio, Italy. *Mem. Soc. Geol. Italiana* **15**, 21–32.

Price, N. B. 1976. Chemical diagenesis in sediments. *Chemical Oceanography* (Eds J. P. Riley & R. Chester) 2nd Edn, Vol. 6, 1–59. New York: Academic Press.

Renz, O. 1936. Stratigraphische and mikropaleontologische untersuchung der Scaglia (Obere Kreide—Tertiar) in Zentralen Apennin, *Eclogae. Geol. Helv.* **29**, 1–149.

Renz, O. 1959. Recherche stratigraphiche e micropaleontologiche sulla Scaglia (Cretaceo supriore—Terziario) dell'Appennino Centrale: Mem. Descr. *Carta Geol. Italia* **29**, 1–173.

Ryan, W. F. B. & Cita, M. G. 1977. Ignorance concerning episodes of oceanwide stagnation. *Marine Geology* **23**, 197–215.

Rieber, H. 1977. Eine Ammonitenfauna aus der oberen Maiolica der Breggia–Schlucht (Tessin Schweiz). *Eclogae Geol. Helv.* **70** (3), 777–788.

Savin, S. M. 1977. The history of the earht's surface temperature during the last 100 million years. *Ann. Rev. Earth Planet. Sci.* **5**, 319–355.

Schlanger, S. O. & Jenkyns, H. C. 1976. Cretaceous oceanic anoxic sediments: causes and consequences. *Geologie en Mijnbouw* **55**, 179–184.

Scientific Staff—Leg 48. 1976. Glomar Challenger sails on Leg 48. *Geotimes* **21**, 19–22.

Scholle, P. A. & Arthur. M. A. 1980. Carbon isotopic fluctuations in pelagic limestones: potential stratigraphic and petroleum exploration tool. *American Association of Petroleum Geologists Bulletin* **64**, 67–89.

Selli, R. 1952. Il bacino del Metauro. *Giorn. Geol.* **21**, 99–125.

Sigal, J. 1977. Essai de zonation du Crétacé Médittérranèen à l'aide de foraminifères planctoniques. *Geol. Mediterran.* **4** (2), 99–103.

Sliter, W. V. 1977. Cretaceous benthic foraminifers from Western South Atlantic Leg 39, Deep Sea Drilling Project. *Initial Reports of Deep Sea Drilling Project* **39**, 657–697.

Sliter, W., Be, A. W. H. & Berger, W. H. (Eds) 1975. Dissolution of deep sea carbonates. *Cushman Found. Foram. Res., Spec. Pub* **13**, 159pp.

Smith, A. G. 1971. Alpine deformation and the oceanic areas of the Tethys, Mediterranean and Atlantic. *Geological Society of America Bulletin* **82**, 2039–2056.

Sorbini, L. 1976. L'Ittiofauna cretacea di cinto Euganeo (Padova, Nord Italia). *Boll. Museo Civico Storia Naturale Verona* **3**, 479–567.

Suess, E. 1979. Mineral phases formed in anoxic sediments by microbial decomposition of organic matter. *Geochimica Cosmochimica Acta* **43**, 339–352.

Summerhayes, C. P. 1981. Organic facies of mid-Cretaceous black shales in the deep North Atlantic. *American Association Petroleum Geologist Bulletin* **65**, 2364–2380.

Thiede, J. & van Andel, T. H. 1977. The paleoenvironment of anaerobic sediments in the Late Mesozoic south Atlantic Ocean. *Earth and Planetary Science Letters* **33**, 301–309.

Thierstein, H. R. 1979. Paleo-oceanographic implications of organic carbon and carbon distribution in Mseozoic deep-sea sediments. *Results of Deep Drilling in the Atlantic Ocean. Second Maurice Ewing Symposium, Am. Geophys, Union* (Eds M. Talwani, W. W. Hayes & W. B. F. Ryan), Vol. 3, 249–274.

Thierstein, H. R. & Berger, W. H. 1978. Injection events in ocean history. *Nature* **276**, 461–466.

Tissot, B., Deroo, G. & Herbin, J. P. 1978. Organic matter in Cretaceous sediments of the North Atlantic. *Results of Deep Sea Drilling in the Atlantic Ocean. Second Maurice Ewing Symposium, Am. Geophys. Union* (Eds M. Talwani, W. W. Hayes & W. B. F. Ryan), Vol. 3, 362–274.

Tissot, B., Durand, B., Espitale, J. & Combaz, A. 1974. Influence of nature and diagenesis of organic matter in formation of petroleum. *American Association Petroleum Geologists Bulletin* **58**, 499–506.

Tissot, B., Demaison, G., Masson, P., Delteil, J. R. & Combaz, A. 1980. Paleonvironment and petroleum potential of Middle Cretaceous black shales in Atlantic Basins. *American Association Petroleum Geologists Bulletin* **64**, 2051–2063.

Vandenberg, J., Klootwijk, C. T. & Wonders, A. A. H. 1978. Late Mesozoic and Cenozoic movements of the Italian Peninsula: Further paleomagnetic data from the Umbrian sequence. *Geological Society of America Bulletin* **89** (1), 133–150.

van Hinte, J. E. 1976. A Cretaceous time scale. *American Association of Petroleum Geologists Bulletin* **60**, 498–156.

von Stackelberg, V. 1971. Facies of sediments of the Indian–Pakistan Continental Margin (Arabian Sea). *Meteor Research Results* **C** (9), 1–73.

Venzo, S. 1954. Stratirafia e tettonica del Flysch (Cretacico-Eocene) del Bergamasco e della Brianza orientale. *Mem. Descr. Carta Geol. d'Italia* **31**, 7–134.

Weissert, H. 1979. Die paleoozeanographie der süwestlichen Tethys in under Unterkreide. *Mitteilungen aus dem Geol. Inst. der ETH und der Univ. Zurich* No. 225, 174pp.

Weissert, H. 1981. The environment of deposition of black shales in the Early Cretaceous. An ongoing controversy. *Soc. Econ. Paleont. Mineral. Spec. Pub.* **32**, 547–560.

Weissert, H., McKenzie, J. & Hochuli, P. 1978. Cyclic anoxic events in the Early Cretaceous Tethys Ocean. *Geology* **7**, 147–151.

Wiedmann, J., Butt, A. & Einsele G. 1978. Vergleich von marokkanischen Kreide-Kustenaufschlussen und Tiefseebohrungen (DSDP): Stratigraphie, Paleoenvironment and Subsidenz an einem passiven Kontinentalrand. *Geol. Rundschau* **67** (2), 454–508.

Winterer, E. L. & Bosellini, A. 1981. Subsidence and sedimentation on Jurassic passive continental margin, Southern Alps, Italy. *American Association of Petroleum Geologists Bulletin* **65** (3), 394–421.

Wonders, A. A. H. 1979. Middle and Late Cretaceous pelagic sediments of the Umbrian Sequence in the Central Apennines. *Proc. Kon. Ned. Ak. Wet., Ser. B.* **82** (2), 171–205.

3. Origin and Geochemistry of Redox Cycles of Jurassic to Eocene Age, Cape Verde Basin (DSDP Site 367), Continental Margin of North-West Africa

W. E. Dean

U.S. Geological Survey, Colorado

J. V. Gardner

U.S. Geological Survey, California

The entire stratigraphic section cored at Deep Sea Drilling Project Site 367 in the Cape Verde Basin is characterized by variations in color and (or) concentration of organic matter in rocks and sediment that range in age from Late Jurassic to Eocene. Late Jurassic (Oxfordian–Kimmeridgian) redox cycles consist of red and green nodular limestones. The Late Jurassic to Early Cretaceous organic carbon-rich facies consists of alternating light-gray limestone and organic carbon-rich olive marlstone in which concentrations of organic carbon are commonly several percent but reach 33%. The Middle to Upper Cretaceous section consists of alternating green and black or green and red shales or claystones, grading into Cretaceous–Tertiary green and red clays, and finally Eocene green and black clays. The youngest sediment (Miocene to Pleistocene) is a carbonate facies that also shows evidence of cyclic interbedding of marl and ooze, but the nature of interbedding has been obscured by drilling disturbance. Most green clays and claystones of all ages have concentrations of organic carbon of less than 0.5% whereas the black clays and shales commonly contain more than 2% organic carbon and have a maximum of 37%.

Many trace elements tend to be enriched in the more reduced lithology in the cyclic redox couplets, i.e. they are more abundant in black or dark-olive colored lithologies than in adjacent light-colored green or red lithologies. Elements that have the greatest variability are V, Cr, and Ni.

The cyclic interbeds of chemically more-reduced and less-reduced, or reduced and oxidized lithologies are interpreted as being mainly the result of cyclic variations in diagenetic redox conditions within the sediments in response to variable rates of supply of organic detritus transported by turbidity currents from shallower water. The accumulation of organic carbon-rich strata is not dependent on having low-oxidation conditions in the bottom waters, but would be aided by restricted bottom-water circulation and low oxygen concentrations. The accumulation of organic carbon-rich strata in the Cape Verde Basin began in the Late Jurassic or the Early Cretaceous and continued intermittently at least into the Eocene. However, the main period of accumulation of organic matter was during the Middle Cretaceous, a period of high organic productivity and decreased oceanic circulation in the world ocean. High concentrations of organic matter of marine, terrestrial or mixed origin accumulated along the shallow continental margins of the North and South Atlantic and was periodically transported to deeper-water environments of sediment accumulation by turbidity currents. In the absence of a high rate of supply of oxygen-consuming organic debris, the bottom waters of the Atlantic Ocean generally contained enough dissolved oxygen to permit the accumulation of oxidized sediments.

1. Introduction

Most of the Jurassic to Pleistocene section cored at Deep Sea Drilling Project (DSDP) Site 367 in the Cape Verde Basin (Figure 1) consists of cyclic interbeds of different colored strata in which the main differences between interbeds are color and (or) the amount of organic matter present (Dean et al., 1978). The main interbedded lithologies are green and black, or green and red clays, claystones, or shales, although the oldest cycles recovered (Lower Jurassic to Lower Cretaceous) are carbonates. The youngest sediments (Miocene to Pleistocene) are also carbonates and probably contain ooze-marl cycles, but they were too severely disturbed by drilling to preserve details of interbedding. In general, the red lithologies contain the lowest concentrations of organic carbon and the black (or dark olive) lithologies contain the highest concentrations. The variable colors clearly reflect redox conditions within the sediments when they were deposited and for some time thereafter, but it is not clear whether the fluctuations in amount of organic matter, related to color, were caused by or were the cause of the variable redox conditions. Nor is it clear whether deoxygenation of bottom waters overlying the sediments was the cause or an effect of fluctuating redox conditions within the sediments.

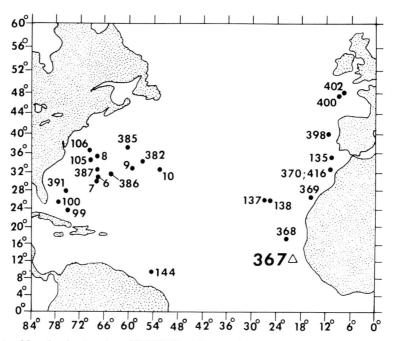

Figure 1.   Map showing location of DSDP Site 367 and other DSDP sites discussed in this report.

DSDP results have shown that strata containing relatively high concentrations of organic matter (more than 2% organic carbon) are common in Middle Cretaceous sections in many parts of the world ocean and particularly in the North and South Atlantic basins (Schlanger & Jenkyns, 1976; Arthur, 1979a; Tucholke & Vogt, 1979; Thierstein, 1979; Arthur & Natland, 1979; Tissot et al., 1979, 1980). These strata are often loosely described as "black shales", although they usually consist of interbedded rocks whose main differences are in color and (or) concentration of organic matter. The most common lithologies are interbedded black or dark greenish-gray shale or claystone and lighter greenish-gray shale or claystone (i.e. interbedded "black" and "green" argillaceous rocks). True black shale usually amounts to less than half of so-called "black shale facies". Commonly, the black beds are finely laminated, whereas the green beds are usually bioturbated.

The purpose of this paper is to describe the organic carbon-rich rocks and sediments that were recovered at Site 367 from the Cape Verde Basin on DSDP Leg 41, and to speculate on how interbedding of chemically more-reduced and less-reduced strata may have originated. This interbedding is characteristic of most mid-Cretaceous sections in the Atlantic. In addition, we present geochemical data to show the effects of fluctuating redox conditions on concentrations of major, minor and trace elements in rocks and sediments. We will concentrate on Site 367 because it contains an excellent Cretaceous section and also provides an opportunity to compare cyclic interbeds of organic carbon-rich and organic carbon-poor strata of Cretaceous age with similar interbeds of Eocene age recovered at the same site.

2.   Redox cycles in sediments and rocks at Site 367

2.1.   Oxfordian–Kimmeridgian red and green nodular limestones

Redox-related cycles in the section recovered at Site 367 begin in Oxfordian and Kimmeridgian strata with alternations of dark reddish-brown and light greenish-gray limestones (Figures 2 and 3). This lithologic unit is identical in age and composition to the Cat Gap Formation of Jansa et al. (1979), recovered at DSDP Sites 99, 100, and 391 in the North American Basin of the western North Atlantic, and the Rosso ad Aptici facies of the classic Jurassic Tethyan sections (Bernoulli, 1972; Bernoulli & Jenkyns, 1974; Jansa et al., 1978, 1979). The red limestone is argillaceous, with common nannofossils, wavy laminations and burrows. The interbedded green limestone consists of boudinage-like structures or coalescing nodules giving the beds wavy, uneven upper and lower contacts. Both red and green limestones are very well cemented, and stylolites are common. This limestone unit was subdivided by Jansa et al. (1978) into radiolarian, filament, and Soccocoma microfacies based on characteristic microfossils present.

We believe that the red and green coloration of the limestones is mainly due to fluctuating diagenetic redox conditions in the sediments in much the same way that red and green cycles in overlying strata are caused by fluctuating redox conditions (see discussion below). The red color is due to coatings of ferric oxides whereas the green color is due to the lack of ferric oxides. Other models involve diagenetic fluctuations in dissolution of aragonite. Bosellini and Winterer (1975) believed that there were diagenetic fluctuations in dissolution of aragonite (largely from am-monites as evidenced from ammonite molds) in response to cyclic fluctuations in the position of the aragonite lysocline. The calcium carbonate was then reprecipitated as light-colored calcite nodules in a red, argillaceous lime mud matrix. Jenkyns (1974) also proposed early diagenetic dissolution of aragonite as the cause of alternating green nodule-rich and red clay-rich layers in the Tethyan Jurassic. However, in Jenkyns' model, the cyclicity was the result of periodic buildup of calcite supersaturation relative to the position of the sediment–water interface.

2.2.   Tithonian–Neocomian black marlstone and white limestone

The Jurassic limestones are overlain by interbeds of laminated dark-olive to black marlstone and bioturbated white to light-gray limestone of Tithonian to Neocomian age (Figures 2, 4 and 5). Upper and lower contacts of the black marlstone beds are

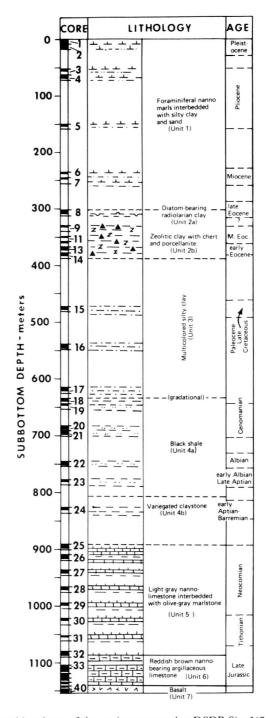

Figure 2.   Stratigraphic column of the section recovered at DSDP Site 367, Cape Verde Basin.

Figure 3. Core photograph of light-green and red nodular limestone of Late Jurassic age, Site 367, Core 34, Section 4.

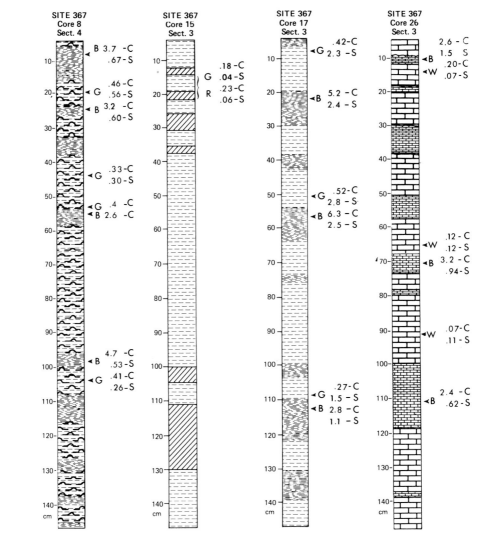

Figure 4. Stratigraphic columns of DSDP Site 367, Core 8, Section 4, Core 15, Section 3, Core 17, Section 3, and Core 26, Section 3. The letter designation by each sample (arrow) is the color of the lithology sampled; B = black; G = green; W = white; R = red. The numbers to the right of the column for Core 15, Section 3 are average concentrations of total carbon (—C) and total sulfur (—S) in three green claystone (G) and three red claystone (R) layers. Numbers to the right of each of the other three columns are percentages of organic carbon (—C) and total sulfur (—S) in samples indicated by arrows.

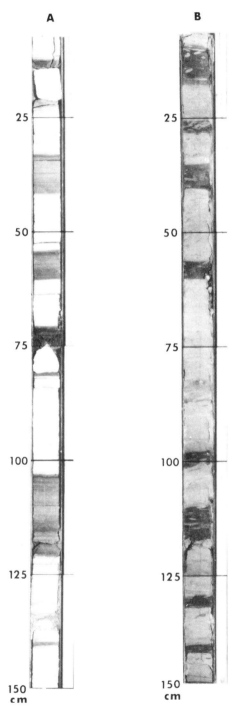

Figure 5.   Core photographs of DSDP Site 367, Core 26, Section 3 (A), and Core 8, Section 4 (B).

usually very sharp (Figure 5). This unit is the equivalent of the Blake-Bahama Formation of Jansa *et al.* (1979) that was recovered at Sites 99, 100, 105, 387 and 391 in the North American Basin; it is the eastern North Atlantic equivalent of the Tithonian to Neocomian white and gray limestone described by Lancelot *et al.* (1972). The unit apparently is equivalent to the Late Tithonian–Barremian Maiolica Formation of Garrison (1967) that outcrops in the Tethyan regions of the Mediterranean (Bernoulli, 1972; Jansa *et al.*, 1979). According to Jansa *et al.* (1979), this unit was deposited in an oxygenated deep bathyal environment. They suggested that the alternating laminated and bioturbated layers may include fine-grained turbidites derived from adjacent carbonate highs. The carbonate content of the laminated limestones and marlstones at Site 367 varies from less than 40% to more than 90% (less than 5% to more than 11% carbonate–carbon). The organic-carbon concentration in the white limestone is usually less than 3% (Figure 6; Table 1). Most of the laminated dark marlstone beds contain more than 5% organic carbon (Table 1), and one bed in Core 26, Section 4 has an organic carbon content of 33% (Dean *et al.*, 1978). The limestone–marlstone couplets in Core 26, where the cycles are particularly well-developed, have an average thickness of about 22 cm. Only crude estimates of sediment accumulation rates can be made because of spot coring and poor fossil preservation, abundance and diversity. An accumulation rate of about 1.2 cm per 10^3 years was estimated by Lancelot, Seibold *et al.* (1978) for the Neocomian limestones and marlstones at Site 367 which gives an average period of about 18 000 years for the carbonate cycles.

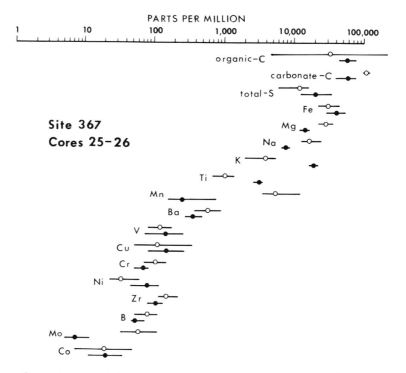

Figure 6.   Comparison of total element concentrations on a carbonate-free basis in 9 samples of white limestone (open circles) and 9 samples of olive-black marlstone (closed circles) from DSDP Site 367, Cores 25 and 26 both of Neocomian age. Circles are located at the geometric mean concentration for each element (Table 1). Bars indicate total observed ranges of element concentration.

Table 1. Geometric mean concentrations of 18 elements on a carbonate-free basis in more organic carbon-rich (black) and less organic carbon-rich (white or green) interbedded lithologies in Cores 25 and 26 (Neocomian), Core 17 (Upper Cretaceous, and Core 8 (Eocene), DSDP Site 367

Element	Method	Cores 25 and 26		Core 17		Core 8	
		white (n=9)	black (n=9)	green (n=5)	black (n=5)	green (n=10)	black (n=8)
% Fe	1	3.1	4.1	5.4	4.7	2.2	2.1
% Mg	1	2.9	1.4	2.0	1.9	2.0	2.7
% Ca	1	36	22	0.37	0.46	0.70	1.0
% Na	1	1.7	0.78	0.83	0.81	1.3	1.5
% K	1	0.40	1.9	2.1	1.9	0.70	0.58
% Ti	1	0.11	0.31	0.41	0.38	0.21	0.23
% Mn	1	0.59	0.027	0.055	0.035	0.014	0.016
% T-C	3	—	—	0.31	5.3	0.35	4.0
% O-C	3	3.2	5.5	—	—	—	—
% C-C	3	10	5.8	—	—	—	—
% T-S	3	1.1	2.0	2.4	2.3	0.25	0.66
ppm B	2	81	57	150	150	82	80
ppm Ba	2	600	370	520	480	350	210
ppm Co	2	21	22	16	16	6	11
ppm Cr	2	104	72	92	120	83	140
ppm Cu	2	107	160	150	160	62	92
ppm Mo	2	62	7.7	7.7	24	1.6	2.9
ppm Ni	2	36	73	71	98	44	90
ppm V	2	130	140	130	270	88	140
ppm Zr	2	160	110	140	160	76	75

(—) means no analyses made. Methods: 1 = x-ray fluorescence; 2 = optical emission spectroscopy; 3 = combustion. T-C = total carbon; O-C = organic carbon; C-C = carbonate carbon; T-S = total sulfur.

2.3. Aptian–Albian red and green claystones

The end of deposition of the Neocomian carbonates was caused by a sudden rise of the carbonate compensation depth (CCD) by more than 2 km (Thierstein, 1979); the remainder of the section through the Eocene is dominated by multicolored, clay-rich strata. The Neocomian carbonates are overlain by interbedded red and green claystones of Late Aptian to Early Albian age. These claystones are composed mainly of quartz, feldspar, smectite, and X-ray-amorphous material, all of which Melieres (1978) interpreted as having been derived from terrigenous sources. These claystones were recovered only in Core 24, but the actual thickness is not known because there are 50 m of uncored interval between this unit and the underlying carbonates in Core 25 and another 50 m uncored interval between this unit and the overlying black shale in Core 23. However, this red and green unit is important because it represents a period of accumulation of more-oxidized, organic carbon-poor sediment between periods of accumulation of more-reduced, organic carbon-rich sediment. This unit was considered by Jansa et al. (1978) to be a subfacies of the overlying black shale facies and does not appear to be present at most DSDP sites in the North Atlantic (Jansa et al., 1978, 1979).

2.4. Aptian/Albian to Turonian black shales

The main period of accumulation of organic carbon in the Cape Verde Basin is represented by a unit of interbedded gray-green claystone and black shale of Late Aptian/Early Albian to Early Turonian age (Figure 2). This unit is equivalent to the

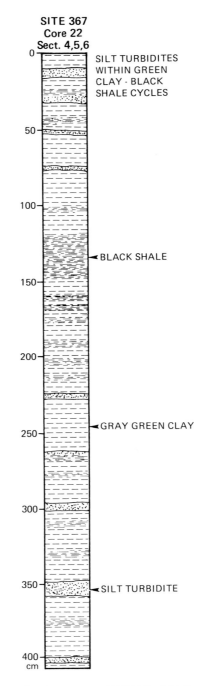

Figure 7. Stratigraphic column of rocks recovermn DSDP Site 367, Core 22, Section 4, 5 and 6 (upper
Albian to lower Turonian).

Hatteras Formation of Jansa *et al.* (1979) in the North American Basin. The mid-Cretaceous black shale facies has been recovered in the western North Atlantic at DSDP Sites 4, 5, 101, 105, 386, 391 and 398, and in the eastern North Atlantic at Sites 135, 137, 138, 367, 368, 398, 400 and 402 (see reviews by Arthur, 1979*a*; Tissot *et al.*, 1979, 1980). In the eastern North Atlantic, the Middle Cretaceous black shale facies is particularly well-developed in the Cape Verde Basin (Site 367) and on Cape Verde Rise (Site 368), where it is intruded by Miocene diabase sills. The unit containing black shale at Site 367 was recovered in Cores 18 through 23 and is at least 150 m thick (636–787 m subbottom; spot coring did not permit clear definition of the upper and lower boundaries of this unit). The black shale actually extends into Core 17 (616–625.5 m subbottom), where it is interbedded with grayish-blue-green silty clay (Figure 4). Siltstone and claystone turbidites are common throughout the black-shale unit, as are interbeds of greenish-gray claystone (Figure 7). The dominant inorganic mineral components in the black shale (quartz, feldspar, smectite and X-ray-amorphous material) are similar to those in the underlying red and green claystones and are exclusively terrigenous (Melieres, 1978). Concentrations of organic carbon in the black-shale beds range from 3% to 37% and are usually more than 10% (Lancelot, Seibold *et al.*, 1978; Figures 4, 7 and 8; Table 1). Concentrations of organic carbon in green claystone interbeds are mostly less than 0.5% (Figures 4 and 8; Table 1).

The cyclic interbeds of green and black shale have periodicities ranging from 20 000 to 10 000 years per cycle with an average of about 50 000 years per cycle (Dean *et al.*, 1978). Cyclic interbeds of oxic–anoxic strata with similar periods have been reported from sections of the Aptian/Albian of Tethyan sequences at Gubbio, Italy, by Arthur & Fischer (1977), in the Aptian to Cenomanian Section at Site 387 on the western Bermuda Rise (McCave, 1979), and in the Albian to Cenomanian section at DSDP Site 398 on Vigo Seamount off the coast of Portugal (Arthur, 1979*a*).

2.5. *Upper Cretaceous–Lower Eocene red and green clay*

The middle Cretaceous black-shale unit in the Cape Verde Basin is overlain by at least 160 m of interbedded grayish-blue-green and reddish-brown clay of Late Cretaceous to Late Paleocene/Early Eocene age (Figures 2 and 4). Like the Aptian/Albian red and green claystones, the Cretaceous/Tertiary green and brown clay section represents a period of accumulation of sediment low in organic matter between strata representing periods of accumulation of sediment rich in organic matter. Couplets of green and red interbeds have an average thickness of about 35 cm. Biostratigraphic control in this part of the section at Site 367 is poor, but accumulation rates are estimated at about 0.6–0.8 cm per 10^3 years, which yield cycle periods of about 50 000 years. This periodicity is about the same as those of cycles in the underlying Middle Cretaceous green and black clays. This unit is equivalent to the Upper Cretaceous–Lower Tertiary multicolored clays at DSDP Site 105 described by Lancelot *et al.* (1972), and the Plantagenet Formation of Jansa *et al.* (1979), recovered at DSDP Sites 7, 9, 105, 382, 385, 386 and 391 in the North American Basin. Jansa *et al.* (1979) interpreted the Plantagenet Formation as having been deposited in an oxygenated, pelagic deep-sea environment similar to the environment presently accumulating pelagic clays in the central North Pacific. At Site 105, this unit is enriched in heavy metals, especially Mn, Zn, Cu, Pb, Cr, Ni and V, which led Lancelot *et al.* (1972) to suggest that the clays were enriched in metals by hydrothermal volcanic exhalations and in this regard were analogous to metal-enriched sediments associated with Red Sea brine deposits or metal-enriched basal sediments overlying basement along the East Pacific Rise. Arthur (1979*b*) proposed that the metals were not derived from volcanic exhalations but from chemically-

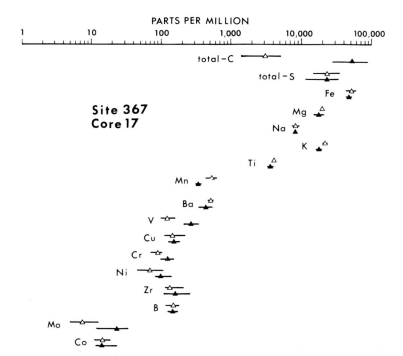

Figure 8. Comparison of total element concentrations in 5 samples of green claystone (open triangles) and 5 samples of black shale (closed triangles) from DSDP Site 367, Core 17 (Upper Cretaceous). Triangles are located at the geometric mean concentration for each element (Table 1). Bars indicate total observed ranges of element concentration.

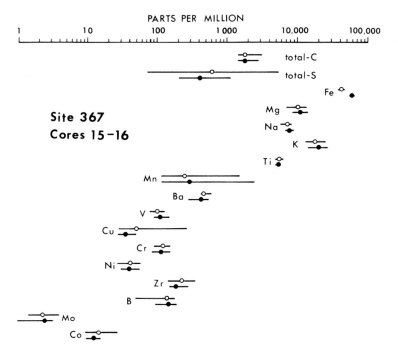

Figure 9. Comparison of total element concentrations in 10 samples of green claystone (open circles) and 10 samples of red claystone (closed circles) from DSDP Site 367, Cores 15 and 16 (Upper Cretaceous). Circles are located at the geometric mean concentration for each element (Table 2). Bars indicate total observed ranges of element concentration.

reduced metallic ions in the anoxic pore waters of the underlying black shales that diffused upward and were oxidized in the slowly accumulating upper Cretaceous clays. The equivalent clays at Site 367, however, are not particularly enriched in any metals, nor are there any significant differences in metal concentrations between red and green clays except for a slightly higher concentration of Fe in the red clays (Figure 9; Table 2). The varicolored claystones at Site 386, the type locality for the Plantagenet Formation of Jansa et al. (1979), consist mainly of zeolites, illite, and montmorillonite. Zeolites are also common in equivalent units at other DSDP sites. Zeolites are not present in the red and green clays at Site 367, however, and the dominant minerals are quartz, smectite, and X-ray-amorphous material (Melieres, 1978).

Table 2.   Geometric mean concentrations of 18 elements in 10 samples each of red and green claystone from Cores 15 and 16 (Upper Cretaceous), DSDP Site 367, Cape Verde Basin

| | Cores 15 and 16 | |
Element	red (n = 10)	green (n = 10)
% Fe	6.0	4.3
% Mg	1.1	0.99
% Ca	0.25	0.28
% Na	0.75	0.71
% K	2.0	1.8
% Ti	0.53	0.54
% Mn	0.029	0.024
% T-C	0.18	0.18
% T-S	0.04	0.06
ppm B	140	130
ppm Ba	420	440
ppm Co	12	14
ppm Cr	110	120
ppm Cu	34	50
ppm Mo	2.3	2.2
ppm Ni	39	41
ppm V	110	100
ppm Zr	190	220

2.6.   Eocene green and black clays

The final episode of accumulation of organic matter in the Cape Verde Basin is represented by at least 85 m of black, zeolitic, radiolarian clays interbedded with greenish-gray, zeolitic, radiolarian clays of Eocene age Figures 2, 4 and 5). The black-clay layers have an average thickness of about 10 cm. The green and black cycles are best developed in Core 8, where they have an average periodicity of about 50 000 years per cycle (Figures 4 and 5; Dean et al., 1978). The lower contact of a black clay layer is always sharp; the upper contact is generally sharp but may be gradational. Bioturbation is common in the green beds but rare in the black beds. The black beds usually contain several percent organic carbon and the green beds contain less than 0.4% organic carbon (Figures 4 and 10; Table 1). The clays in Cores 9 through 14 contain abundant zeolites (mostly clinoptilolite) and interbedded chert and porcellanite. The dominant clay minerals are attapulgite and sepiolite which Melieres (1978) interpreted to have spilled over from restricted coastal basins where

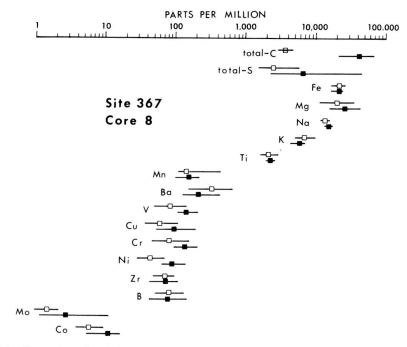

Figure 10.   Comparison of total element concentrations in 10 samples of green clay (open squares) and 8
      samples of black clay (closed squares) from DSDP Site 367, Core 8 (Eocene). Squares are located at
      the geometric mean concentration for each element (Table 1). Bars indicate total observed ranges of
      element concentration.

they formed from the solubilized products of weathered continental rocks. These
Eocene clays are equivalent to the lower and middle Eocene Bermuda Rise
Formation of Jansa *et al.* (1979), recovered at DSDP Sites 6, 7, 8, 9, 10, 106, 385, 386
and 387 in the western North Atlantic. Lithologies of this formation include
radiolarian claystone, mud, and ooze; nannofossil claystone; zeolitic clay; and silty
mudstone; all of these lithologies contain interbedded chert (Jansa *et al.*, 1979). At
the type locality, Site 387 on the Bermuda Rise, the formation consists of cyclic
interbeds of chert, silicified radiolarian mudstone and claystone, and calcareous
mudstone interpreted to be sequences of turbidites (Jansa *et al.*, 1979; McCave, 1979).

3.   Geochemistry

3.1.   Methods

Couplets of green and black lithologies from Core 8 (Eocene), Core 17 (Upper
Cretaceous), and Core 19 (Middle Cretaceous, Turonian); white and black
lithologies from Cores 25 and 26 (Middle Cretaceous, Barremian to Aptian), and
green and red lithologies from Cores 15 and 16 (Upper Cretaceous) were analyzed for
26 major, minor and trace elements by a combination of X-ray fluorescence, optical
emission spectroscopy, atomic absorption, and combustion techniques. Details of
the analytical methods are described in the analytical sections of Miesch (1976). Ten
of seventy total samples were selected at random for duplicate analyses, and all eighty
analytical samples were submitted in a random sequence to the analytical laboratories of the U.S. Geological Survey.

3.2. *Results*

Geometric mean concentrations of 18 elements are listed in Table 1 for each lithologic type from cyclic interbeds in Cores 8, 17, 25 and 26, and in Table 2 for cyclic interbeds of red and green claystone from Cores 15 and 16. Concentrations of all elements except Ca, total-C, and carbonate-C in samples from Cores 25 and 26 are expressed on a carbonate-free basis; samples from all other cores did not contain any carbonate and concentrations are as measured in the bulk sample. The 18 elements listed are the ones that were most variable between couplet lithologies. Other elements analyzed for but not included in Tables 1 and 2 are Si, Al, Ga, La, Nb, Sc, Sr, and Y. Bar plots showing total ranges and geometric means of concentrations of 17 of these 18 elements (Ca is not included in the plots) in couplet lithologies from Cores 25 + 26, 17, 15 + 16, and 8 are shown in Figures 6, 8, 9 and 10 respectively.

Table 1 and Figures 6, 8 and 10 show that some elements tend to be enriched in the more reduced lithologies in cycle couplets (e.g. black or dark-olive lithologies). Concentrations of V and Ni are higher in the more-reduced lithologies in all couplets sampled. In most couplets, concentrations of Cr, Cu and Mo are higher in the more-reduced lithologies, although the carbonate-free concentrations of Cr and Mo are higher in the light-gray limestones than in the dark-olive marlstones in Cores 25 and 26. The considerably lower concentration of Mo in the olive-black marlstones in Cores 25 and 26, relative to the white limestones, is somewhat surprising although Mo can be either concentrated or depleted in anoxic strata (e.g. Bertine, 1972; Bertine & Turekian, 1973; Berrang & Grill, 1974; Lahann, 1977). Molybdenum is easily remobilized under reducing conditions and may diffuse from the sediments into the overlying water (Contreras *et al.*, 1978; Holland, 1979). However, Mo also may be coprecipitated with ferrous sulfide under anoxic conditions, resulting in relatively high concentrations of Mo in some anoxic sediments (Berrang and Grill, 1974). Adsorption of Mo by MnO_2 is probably the most important process removing Mo under oxidizing conditions and may be the explanation for the higher concentration of Mo in the white limestone samples from Cores 25 and 26.

The differences in total sulfur concentrations between organic carbon-rich and organic carbon-poor lithologies (shown in Figures 6, 8 and 10 and Table 1) presumably are due to variations in pyrite. X-ray diffraction results by Melieres (1978) show that detectable amounts of pyrite occur only in Cores 17 through 24 (Upper Aptian–Lower Albian to Upper Cretaceous), although X-ray diffraction may detect not less than several percent pyrite. The high concentrations of total sulfur in both green and black lithologies in Cores 17 (Table 1) and 19 therefore reflect the presence of higher concentrations of pyrite in these cores than in the other cores from which samples were analyzed.

The enrichment of certain trace elements—especially Cu, Zn, Mo, V, Ni, Cr, Ba, and Pb—in sediment rich in organic matter has been reported by many investigators (e.g. Tourtelot, 1964; Wedepohl, 1964; Brongersma-Sanders, 1965; Calvert & Price, 1970; Vine & Tourtelot, 1969, 1970; Volkov & Fomina, 1974; Chester *et al.*, 1978; Leinen & Stakes, 1979, among others). The association of high trace-element concentrations with organic matter may be the result of concentration of these elements by organisms (bioconcentration) or by chemical sorption and precipitation processes. The known bioconcentration by marine plankton of a number of trace elements, especially Pb, Ni, Cu, Zn, Mn, and Fe (e.g. Vinogradov, 1953; Goldberg, 1957; Boyle & Lynch, 1968; Martin & Knauer, 1973; Leinen & Stakes, 1979) suggests that bioconcentration is a potentially important mechanism for incorporation of certain trace elements in organic carbon-rich marine sediments. However, the great effectiveness of adsorption of trace metals by clay minerals and organic matter and coprecipitation of trace metals, particularly as sulfide minerals, suggests that these processes may play a more important role in the removal of trace metals from sea water (e.g. Tourtelot, 1964, 1979; Holland, 1979, among others). Unfortunately, it is not possible to separate the effects of bioconcentration and chemical processes, especially in anoxic, organic carbon-rich strata.

Tourtelot (1964) studied the minor-element compositions of coeval Upper Cretaceous marine and non-marine shales of the western interior of the United States and found that element concentrations were consistently higher in the marine shales. From this observation, he concluded that sorption and syngenetic mineral formation in the marine environment were important, if not the most important, processes affecting the minor-element content of shales. The final minor-element composition is thus the result of cycles of sorption–desorption and (or) precipitation–dissolution as the sediments are subjected to varying chemical conditions in depositional and diagenetic environments. The greatest control is probably exerted by the redox conditions during diagenesis.

Vine & Tourtelot (1970) also concluded that sorption processes are important in metal enrichment in black shales but felt that some prior concentration of metals from seawater (e.g. by bioconcentration) was required. They stated that the availability of metals in the various solutions that the organic matter comes in contact with throughout its history is probably the most significant factor in determining metal enrichment, and that interstitial pore fluids are probably the most important of these solutions. If thin beds of black shale are interstratified with other lithologies, these beds may become sinks for elements mobilized and transported in pore waters during compaction.

4. Discussion—origin of redox cycles

Interbedding of sediments alternately rich and poor in organic matter can be the result of either differential preservation of organic matter, differential rates of supply of organic matter, and (or) differential sedimentation rates. Differential preservation can result if bottom waters at the site of accumulation are alternately oxic and anoxic (or near-anoxic). Cyclic variation in the amount of organic matter accumulating at a continental margin site may be explained by cyclic fluctuations in the thickness and intensity of a midwater oxygen-minimum layer. Both depositional models (basin deoxygenation and expansion and intensification of an oxygen-minimum layer) have been proposed to explain the accumulation of organic carbon-rich strata (e.g. Schlanger & Jenkyns, 1976; Ryan & Cita, 1977; Fischer & Arthur, 1977; Thiede & van Andel, 1977; Arthur & Schlanger, 1979). Both models require low to zero concentrations of dissolved oxygen in part of the water column as a result of reduced advection of oxygenated water and (or) increased supply of organic matter, and both models imply that reducing conditions in the sediments (and therefore the increased degree of preservation of organic matter) are the *result* of anoxic or near-anoxic conditions in the overlying waters.

Black, laminated, organic carbon-rich Cretaceous strata are so widespread in the Atlantic that the closed-basin model has been generally accepted as the most likely analog for the Early to Middle Cretaceous South and North Atlantic Oceans. We find it difficult, however, to explain the interbedding of chemically more reduced and less reduced strata by fluctuating deoxygenation of bottom waters. Most of the Jurassic and Cretaceous strata in the Atlantic contain abundant evidence that they were deposited in bioturbated, well-oxygenated bottom-water environments; only the interbedded dark-colored, organic carbon-rich lithologies even suggest anaerobic depositional conditions. For example, most of the carbonates in the Upper Jurassic section in the North Atlantic were deposited in well-oxygenated, deep-basin environments interrupted periodically by deposition of finely laminated, dark-olive, organic carbon-rich layers. Surely these dark layers do not represent periods of complete basin stagnation. Interbedding of anoxic-oxic strata is very well developed in the Cape Verde Basin but is certainly not unique to that area. This interbedding at Site 367 began in the Jurassic and continued at least into the Eocene and probably into the Pleistocene. If the accumulation of high concentrations of organic matter is related to complete basin stagnation, then how is this stagnation created and destroyed with great frequency and regularity over such a long period of time? It is not difficult to imagine abrupt changes in bottom-water circulation in a restricted basin such as the Baltic Sea, the Mediterranean Sea, or even the Early Cretaceous Atlantic Ocean, but it is difficult to imagine total stagnation of the Atlantic during the Late Cretaceous and Tertiary.

We favor a third depositional model in which varying redox conditions in the sediments are caused mainly by post-depositional variations in oxygen demand within the sediments, largely in response to the amount of organic detritus. The organic detritus may be from autochthonous marine planktonic production, or from terrestrial organic debris. According to Tissot et al. (1979, 1980), pure marine organic matter (i.e. organic matter containing type II kerogen) in continuous sections of Lower Cretaceous rocks in the North Atlantic is found only off the coast of north-west Africa at DSDP Sites 367 and 369. Everywhere else, terrestrial organic matter with some admixtures of highly degraded, residual organic matter dominates the Lower Cretaceous organic carbon-rich strata. Terrestrial organic matter is also dominant in Upper Cretaceous sections at most sites in the North Atlantic except in the Cape Verde Basin (Site 367), on Cape Verde Rise (Site 368), and on Demerara Rise (Site 144), where marine organic matter is dominant (Tissot et al., 1979, 1980). It therefore appears that the Cretaceous organic carbon-rich strata in the south-eastern North Atlantic, and at Site 367 in particular, have more in common with similar strata in the South Atlantic where most of the organic matter is derived from marine sources than with equivalent strata in the rest of the North Atlantic that are dominated by terrestrial organic matter (Tissot et al., 1979, 1980). This is not surprising because during this time the continental margin of west Africa was within a zone of arid tropical climate with intense upwelling offshore and associated high biological productivity, whereas the rest of the continental margin of the North Atlantic was within humid, temperate to subtropical climates without intense upwelling. In spite of these differences in sources of organic matter, most of the organic carbon-rich strata are interbedded with bioturbated, organic carbon-poor strata.

The problem, then, is to provide a mechanism that will produce a pulsing of organic matter from a variety of sources within the same general Early to Middle Cretaceous time, and over large areas of the North and South Atlantic. We suggest that turbidity currents are the most likely mechanisms for periodic emplacement of sediments rich in organic matter into a deep abyssal environment that was more or less oxygenated. Such a mechanism was proposed by Jansa et al. (1979) for deposition of Neocomian olive-black, laminated marlstones within white, bitur-bated limestones in the North Atlantic (Figures 4 and 5). Cyclic alternations of marlstone and carbonaceous claystone containing mostly terrestrial organic matter of Aptian/Albian age at DSDP Site 400 (Bay of Biscay) were interpreted by de Graciansky et al. (1979) as the results of deposition of mud turbidites onto a deep abyssal plain. Aptian/Albian black shales at Site 402, just upslope from Site 400 in the Bay of Biscay, are also turbidites deposited at the front of two prograding shelves (de Graciansky et al., 1979). At Site 398 on Vigo Seamount, Upper Barremian to Aptian black shales and Aptian/Albian black shales contain graded siltstone and sandstone and are interpreted to have been emplaced by turbidity currents on the

outer part of a deep-sea fan (de Graciansky & Chenet, 1979). The abundance of
terrigenous material at Site 398, including sand, silt, and land-plant-derived organic
matter, was also pointed out by Arthur (1979*a*). Coarse-clastic turbidites are also
associated with Middle Cretaceous rhythmic black shales at Sites 386 and 387 (de
Graciansky *et al.*, 1979). The turbidite nature of the rhythmic organic carbon-rich
strata at Sites 386 and 387, and also at Site 391 in the Blake-Bahama Basin, was
confirmed by personal observations by D. A. V. Stow and one of us (WED), who
recently examined all organic carbon-rich Middle Cretaceous strata recovered from
DSDP sites in the North Atlantic through Leg 48. Silt-size turbidites occur at Site
367 in the Cape Verde Basin within the cycles of green claystone and black shale of
Aptian/Albian to Turonian age in Cores 19 through 23 (Dean *et al.*, 1978). Site 370
encountered a submarine fan in a deep basin beneath the lower continental rise off
Morocco and was redrilled as Site 416. Most of the recovered section at these two
sites consists of turbidites. The Aptian to Cenomanian section at Site 370 consists of
dark greenish-gray to olive-black clay or claystone. The concentration of organic
carbon is usually between 0.5 and 1.0% but was measured as high as 4.8% (Dean *et
al.*, 1978). There appears to be abundant evidence of turbidity-current deposition
within the middle Cretaceous "black shales" in the North Atlantic.

The best example of turbidity-current control on the deposition of cyclic
interbeds of organic carbon-rich deep-sea sediment that we know of comes from the
South Atlantic at Site 530 in the southern Angola Basin, recently cored on DSDP
Leg 75 (Dean *et al.*, 1981; Hay, Sibuet *et al.*, in press). Here, 260 beds of black shale
as much as 62 cm thick (average of 4.3 cm) and containing 1.4–16% organic carbon
are superimposed on a dominant section of Middle to Upper Cretaceous red and
green claystone deposited by a combination of pelagic, hemipelagic, and turbiditic
processes in oxygenated bottom water. The beds of black shale comprise about 10%
of the recovered section just above Late Albian basement, reach a maximum of about
50% of the recovered section 65–70 m above basement, and end 160 m above
basement, although the red and green claystones continue for another 110 m of
section. Beds of green claystone commonly occur as reduction "halos" on either side
of a black shale bed. A green-black-green reduced "sandwich" may occur alone
within a predominant oxidized red claystone, or several closely-spaced, reduced
sequences may merge to form interbedding of black and green lithologies. The
common interbedding of red, green and black layers, and the bioturbation of much of
the sediment but its absence in about half the black shale beds suggests that there was
a delicate balance between oxidizing and reducing conditions in the Angola Basin
and sediments at this time. The section above the black-shale-bearing unit consis
of interbedded red and green claystone where the green beds are also the result of
reduction of the red claystone, usually on either side of a thin silt lamina or layer. The
sequence in the Angola Basin at Site 530 suggests that the deposition of oxidized red
clay low in organic matter in an oxygenated bottom-water environment was
periodically interrupted by the deposition of clay containing abundant organic
matter of marine origin that resulted in reduction of iron and color changes within
the sediment.

Turbidity-current deposition, therefore, is a possible mechanism for producing
cyclic interbedding of red, green and black lithologies in which the color variations
are the result of diagenetic redox changes within the sediment. Figure 11 summarizes
our concept of how variations in rate of supply of organic matter, from any source,
and of marine, terrestrial, or mixed origin, can result in cyclic red, green, and black
interbeds. Similar environmental interpretations of facies in the Liassic of
Yorkshire, England, were made by Morris (1979) and Demaison & Moore (1980). In
the environment illustrated in Figure 11(A), the rate of supply of organic matter is
low, the bottom waters are more or less well oxygenated, and red clay accumulates.

Iron would be present mainly as ferric oxides and hydroxides and sulfur would be present mainly as SO_4^{2-}. In the environment illustrated in Figure 11(B), the rate of supply of organic matter is higher (indicated diagrammatically by the greater number of black arrows), and the bottom waters are still more or less well oxygenated. However, increased oxygen demand in the sediment, mainly from the decomposition of the increased organic matter, has resulted in a zero-oxygen isopleth at or near the sediment-water interface, and the sediment is reduced. Iron would be mainly present as Fe^{2+} and ferrous sulfide (FeS), and sulfur would be present as SO_4^{2-}, H_2S, HS^-, and FeS. In the environment illustrated in Figure 11(C), the rate of accumulation of organic matter is so great that the sediment becomes completely anoxic and reduced, the zero-oxygen isopleth may be some distance above the sediment-water interface, and H_2S may be present in the bottom waters. The extent of deoxygenation of the bottom waters would be controlled by the

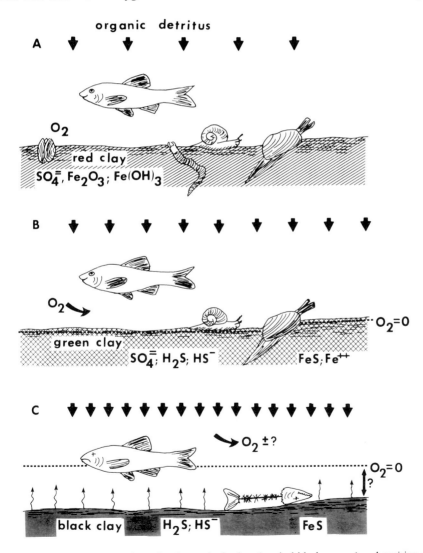

Figure 11.   Variations in rate of supply of organic detritus (vertical black arrows) and position of the boundary between oxidizing and reducing conditions ($O_2 = O$ isopleth) in deep-sea environments. See text for discussion.

rate of oxygen consumption, rate of H_2S production, rate of oxygen supply to the bottom waters, strength of bottom currents, and rate of sedimentation. Iron would be present as ferrous sulfide (plus some Fe^{2+}); sulfur would be present mainly as H_2S, HS^-, and FeS; and the benthic epi- and infauna would be eliminated. The black color would be mainly due to organic matter and FeS. The FeS eventually would be converted to FeS_2 (pyrite or marcasite), which would tend to lighten the color of the sediment. However, maturation of the organic matter would tend to darken the color of the sediment with the result that the black mud would probably become black or at least dark-gray shale through diagenesis. If the initial deposition resulted in interbedding of layers more and less enriched in organic matter but otherwise homogeneous, then diffusion of SO_4^{2-} into layers enriched in organic matter to replace SO_4^{2-} consumed by sulfate reduction may produce localized concentrations of FeS and an intensification of color differences (Berner, 1969).

Turbidity currents, then, are a mechanism that would provide variable amounts of organic matter of variable composition, and would provide rapid burial of the organic matter that would improve its preservation. Why was this process so frequent within the middle Cretaceous? As we have seen at Site 367 in the Cape Verde Basin, accumulation of organic carbon-rich strata actually began in the Early Cretaceous or Late Jurassic and continued at least into the Eocene, but the main period of accumulation of organic matter was during the Middle Cretaceous. Variations in the amount of organic matter supplied to one site, even over a more or less restricted time period, can be explained by local pockets of accumulation of organic matter of marine or terrestrial origin in shallower water. This organic matter is then periodically transported by turbidity currents to deeper-water environments of sediment accumulation. However, local pockets of accumulation of organic matter (such as Walvis Bay, south-western Africa, or Santa Barbara Basin, southern California, today) are the result of local circulation patterns, and it is unlikely that such conditions could have existed around the entire margin of the Middle Cretaceous Atlantic Ocean as well as on isolated seamounts and plateaus in the Pacific Ocean without the aid of some world-wide conditions of ocean circulation and productivity. The Early to Middle Cretaceous apparently was a period of warm surface- and bottom-water temperatures; of maximum extent of warm, shallow inland and marginal seas; and of more sluggish surface- and bottom-water circulation, perhaps enhanced by salinity stratification in a much narrower proto-Atlantic Ocean (Schlanger & Jenkyns, 1976; Fischer & Arthur, 1977; Thiede & van Andel, 1977; Arthur & Schlanger, 1979, among others). It is, therefore, not surprising that large amounts of terrestrial and marine organic matter were produced and accumulated during this time. Most of the organic matter that accumulated in the western and north-eastern North Atlantic was terrestrial, derived from humid temperate regions of Europe and North America. Most of the organic matter that accumulated in the South Atlantic and south-eastern North Atlantic was marine, probably because of lack of dilution by terrestrial organic matter from the more arid African continent. In the Pacific Ocean, high concentrations of predominantly marine organic matter accumulated during the general Middle Cretaceous time only on isolated seamounts and plateaus as a result of coincidence of local structural and circulation factors with a period of high productivity (Schlanger & Jenkyns, 1976; Dean *et al.*, 1981; Thiede *et al.*, this volume).

Periods of accumulation of organic matter in the Atlantic Ocean during the Cretaceous varied in duration and intensity. Two periods of maximum accumulation of organic matter are generally recognized; one in the Late Aptian to Early Albian and one in the Late Cenomanian to Early Turonian (Schlanger & Jenkyns, 1976;

Arthur, 1979*a*; Tissot *et al.*, 1980). These periods probably represent a coincidence of several factors acting to produce and preserve organic matter. Prior to the deposition of a lithologic unit rich in organic matter (e.g. Early Aptian), bottom waters contained abundant dissolved oxygen, and sediment that was transported to the site of deposition had a low oxygen demand so that oxidized (red) sediment accumulated and was preserved. During periods when organic carbon-rich strata were deposited (e.g. Aptian–Albian), higher rates of production of marine organic matter and (or) higher rates of supply of terrestrial organic matter would cause the expansion and intensification of a midwater oxygen minimum. Both the increased rate of supply of organic matter and the expanded and more intense oxygen minimum would increase the amount of organic matter accumulating on the bottom both in terms of area covered by organic carbon-rich sediment and amount of organic matter per unit area. Consequently, more of the sediment transported from shallower sites to deeper sites of accumulation would have a high oxygen demand and result in reduced (green) sediment. Under extreme conditions, black sediment would accumulate and anoxic or near-anoxic conditions might extend to abyssal depths. Any other factor that would affect the concentration of dissolved oxygen, such as warmer temperatures, increased salinity stratification, or more sluggish circulation, would aid in bottom-water deoxygenation as postulated by Schlanger & Jenkyns (1976) and Fischer & Arthur (1977).

If a black, organic carbon-rich bed is more or less isolated within oxidized or less-reduced, highly bioturbated strata, the color variations may be entirely diagenetic and the bottom waters may have been oxygenated at all times. We suggest that examples of strata formed under these conditions are the striking white and black Neocomian limestones and marlstones (Figure 4, core 26 sect. 3) and the green and black Eocene clays (Figure 4, core 26 sect. 4) at Site 367. Under extreme conditions of production and accumulation of organic matter (e.g. during the Aptian–Albian or Cenomanian–Turonian), the rate of supply of oxygen-consuming sediment may have been so great and the rate of supply of oxygen-bearing surface water may have been so slow that bottom waters may have become anoxic. At such times the bottom waters may have contained barely enough oxygen to support a benthic fauna and were so poised that relatively small changes in flux of organic matter and (or) circulation may have caused anoxia or near-anoxia. The main point that we wish to make, however, is that we do not believe that an organic carbon-rich bed between organic carbon-poor beds is the *result* of anoxic bottom water, but rather is the result of an increase in the relative amount of organic matter in the sediment being deposited. Bottom-water and mid-water anoxia may have occurred during periods of accumulation or organic carbon-rich strata but it is not a necessary *cause* of these strata.

The periodicities of all of the redox cycles described in this paper are similar to periodicities of carbonate dissolution, dilution, sea-surface temperature fluctuations, and other parameters that are correlated with climatic changes in the late Neogene and Quaternary. Possibly, the cyclic patterns of sedimentation recorded in most of the section at Site 367 and elsewhere in the Atlantic Ocean were responding to climatic forcing similar to Pleistocene analogs. We suggest that the supply of organic detritus was pulsing in response to climate change. This pulsing of organic matter extended throughout the history of the Atlantic Ocean with varying intensity. During several periods of time (e.g. the Aptian–Albian and Cenomanian–Turonian; the Oceanic Anoxic Events of Schlanger & Jenkyns, 1976) the pulsating input of organic matter because of a coincidence of several climatic and oceanographic factors apparently caused reduced oxygen conditions in the bottom waters.

Acknowledgments

This paper benefited greatly from very helpful discussions with M. A. Arthur, H. A. Tourtelot, W. W. Hay and D. A. V. Stow. We greatly appreciate the help of the DSDP East Coast Repository staff in assisting with the examination of numerous cores containing Middle Cretaceous organic carbon-rich strata. We thank M. A. Arthur, H. A. Tourtelot and J. S. Leventhal for critical reviews of the manuscript, and Louise Reif for massaging its several drafts through the word processor.

References

Arthur, M. A. 1979a. North Atlantic Cretaceous black shales: the record at Site 398 and a brief comparison with other occurrences. *Initial Reports of the Deep Sea Drilling Project* **47** (2), 719–751.

Arthur, M. A. 1979b. Origin of Upper Cretaceous multicolored claystones of the western Atlantic. *Initial Reports of the Deep Sea Drilling Project* **43**, 417–420.

Arthur, M. A. & Fischer, A. G. 1977. Upper Cretaceous–Paleocene magnetic stratigraphy at Gubbio, Italy: I. Lithostratigraphy and sedimentology. *Geological Society of America Bulletin* **88**, 367–389.

Arthur, M. A. & Natland, J. H. 1979. Carbonaceous sediments in the North and South Atlantic: the role of salinity in stable stratification of early Cretaceous basins. *Deep Drilling Results in the Atlantic Ocean: Continental Margins and Paleoenvironment* (Eds M. Talwani, W. W. Hay, and W. B. F. Ryan), Vol. 3, 375–401. Washington, American Geophysical Union, Maurice Ewing Series.

Arthur, M. A. & Schlanger, S. O. 1979. Cretaceous "oceanic anoxic events" as causal factors in development of reef-reservoired giant oil fields. *American Association of Petroleum Geologists Bulletin* **63**, 870–885.

Berner, R. A. 1969. Migration of iron and sulfur within anaerobic sediments during early diagenesis. *American Journal of Science* **267**, 19–42.

Bernoulli, D. 1972. North Atlantic and Mediterranean Mesozoic facies: a comparison. *Initial Reports of the Deep Sea Drilling Project* **11**, 801–871.

Bernoulli, D. & Jenkyns, H. C. 1974. Alpine, Mediterranean, and central Atlantic Mesozoic facies in relation to the early evolution of the Tethys. *Modern and Ancient Geosynclinal Sedimentation: Soc. Economic Paleontologists and Mineralogists Special Publication 19* (Eds R. H. Dott, Jr & R. H. Shaver), 129–160.

Berrang, P. G. & Grill, E. V. 1974. The effect of manganese oxide scavenging on molybdenum in Saanich Inlet, British Columbia. *Marine Chemistry* **2**, 125–148.

Bertine, K. K. 1972. The deposition of molybdenum in anoxic waters. *Marine Chemistry* **1**, 43–53.

Bertine, K. K. & Turekian, K. K. 1973. Molybdenum in marine deposits. *Geochimie et Cosmochimie Acta* **37**, 1415–1434.

Bosellini, A. & Winterer, E. L. 1975. Pelagic limestone and radiolarite of the Tethyan Mesozoic: a genetic model. *Geology* **3**, 279–282.

Boyle, R. W. & Lynch, J. J. 1968. Speculation on the source of zinc, cadmium, lead, copper, and sulfur in Mississippi Valley and similar types of lead-zinc deposits. *Econ. Geology* **63**, 421–422.

Brongersma-Sanders, M. 1965. Metals of Kupferschiefer supplied by normal seawater. *Geol. Rundschau* **55**, 365–375.

Calvert, S. E. & Price, N. B. 1970. Minor metal contents of recent organic-rich sediment off South West Africa. *Nature* **227**, 593–595.

Chester, R., Griffiths, A. & Stoner, J. H. 1978. Minor metal content of surface seawater particulates and organic-rich shelf sediments. *Nature* **275**, 308–309.

Contreas, R., Fogg, T. R., Chasteen, N. D., Gaudetter, H. E. & Lyons, W. B., 1978. Molybdenum in pore waters of anoxic marine sediments by electron paramagnetic resonance spectroscopy. *Marine Chemistry* **6**, 365–373.

Dean, W. E. & DSDP Leg 75 Shipboard Scientific Party. 1981. Cretaceous black-shale deposition within an oxidized red-clay, turbidite environment, southern Angola Basin, South Atlantic Ocean (abs.). *American Association of Petroleum Geologists Bulletin* **65**, 917.

Dean, W. E., Gardner, J. V., Jansa, L. F., Cepek, P. & Seibold, E. 1978. Cyclic sedimentation along the continental margin of northwest Africa. *Initial Reports of the Deep Sea Drilling Project* **41**, 965–989.

Dean, W. E., Claypool, G. E. & Thiede, J. 1981. Origin of organic carbon-rich mid-Cretaceous limestones, Mid-Pacific Mountains and southern Hess-Rise. *Initial Reports of the Deep Sea Drilling Project* **62**, 877–890.

de Graciansky, P. C., Auffret, G. A., Dupeuble, P., Montadert, L. & Müller, C. 1979. Interpretation of depositional environments of the Aptian/Albian black shales on the north margin of the Bay of Biscay (DSDP Sites 400 and 402). *Initial Reports of the Deep Sea Drilling Project* **43**, 877–907.

de Graciansky, P. C. & Chenet, P. Y. 1979. Sedimentological study of Cores 138 to 56 (upper Hauterivian to middle Cenomanian): an attempt at reconstruction of paleoenvironments. *Initial Reports of the Deep Sea Drilling Project* **47**, 403–418.

Demaison, G. J. & Moore, G. T. 1980. Anoxic environments and oil source bed genesis. *American Association of Petroleum Geologists Bulletin* **64**, 1179–1209.

Fischer, A. G. & Arthur, M. A. 1977. Secular variations in the pelagic realm. *Deep Water Carbonate Environments: Society of Economic Paleontologists and Mineralogists Special Publication 25* (Eds H. E. Cook & P. Enos), 19–50.

Goldberg, E. D. 1957. Biogeochemistry of trace metals. *Treatise on Marine Ecology and Paleoecology* (Ed. J. W. Hedgepeth), vol. 1, Ecology: Geological Society, 345–357.

Hay, W. W., Sibuet, J. C. & DSDP Leg 75 Shipboard Scientific Party. In press. Sedimentation and accumulation of organic carbon in the Angola Basin and on Walvis Ridge: preliminary results of Deep Sea Drilling Project Leg 75. *Geological Society of America Bulletin.*

Holland, H. D. 1979. Metals in black shales—a reassessment. *Economic Geology* **74**, 1676–1680.

Jansa, L. F., Gardner, J. V. & Dean, W. E. 1978. Mesozoic sequences of the central North Atlantic. *Initial Reports of the Deep Sea Drilling Project* **41**, 991–1031.

Jansa, L. F., Enos, P., Tucholke, B. E., Gradstein, F. M. & Sheridan, R. E. 1979. Mesozoic–Cenozoic sedimentary formations of the North Atlantic basin; western North Atlantic. *Deep Drilling Results in the Atlantic Ocean: Continental Margins and Paleoenvironment* (Eds M. Talwani, W. W. Hay & W. B. F. Ryan), Washington: American Geophysical Union, Maurice Ewing Series, Vol. 3, 1–57.

Jenkyns, H. C. 1974. Origin of red nodular limestones (Ammonitico Rosso, Knollenkalke) in the Mediterranean Jurassic: a diagenetic model. *Pelagic Sediments: On Land and Under the Sea: International Association of Sedimentologists, Special Publication 1* (Eds K. J. Hsü & H. C. Jenkyns), 249–271.

Lahann, R. W. 1977. Molybdenum and iron behaviour in oxic and anoxic lake waters. *Chemical Geology* **20**, 315–323.

Lancelot, Y., Hathaway, J. C. & Hollister, C. D. 1972. Lithology of sediments from the western North Atlantic, Leg 11, Deep Sea Drilling Project. *Initial Reports of the Deep Sea Drilling Project* **11**, 901–950.

Lancelot, Y., Seibold, E. *et al.* 1978. Site 367: Cape Verde Basin. *Initial Reports of the Deep Sea Drilling Project* **41**, 163–231.

Leinen, M. & Stakes, D. 1979. Metal accumulation rates in the central equatorial Pacific during Cenozoic time. *Geological Society of America Bulletin* **90**, 357–375.

Martin, J. H. & Knauer, G. A. 1973. The elemental composition of plankton. *Geochim. et Cosmochim. Acta* **37**, 1639–1653.

McCave, I. N. 1979. Depositional features of organic black and green mudstones at DSDP Sites 386 and 387, western North Atlantic. *Initial Reports of the Deep Sea Drilling Project* **43**, 411–416.

Melieres, F. 1978. X-ray mineralogy studies, Leg 41, Deep Sea Drilling Project, eastern North Atlantic Ocean. *Initial Reports of the Deep Sea Drilling Project* **41**, 1065–1086.

Miesch, A. T. 1976. Geochemical survey of Missouri—methods of sampling, laboratory analysis, and statistical reduction of data. *U.S. Geological Survey Prof. Paper 954-A*, 39pp.

Morris, K. A. 1979. A classification of Jurassic marine shale sequences: an example from the Toarcian (Lower Jurassic) of Great Britain. *Paleogeography, Paleoclimatology and Paleoecology* **26**, 117–120.

Ryan, W. B. F. & Cita, M. B. 1977. Ignorance concerning episodes of oceanwide stagnation. *Marine Geology* **23**, 197–215.

Schlanger, S. O. & Jenkyns, H. C. 1976. Cretaceous oceanic anoxic events: causes and consequences. *Geologie en Mijnbouw* **55**, 179–184.

Thiede, J., Dean, W. E. & Claypool, G. E. 1983. Oxygen deficient depositional paleoenvironments in the mid-Cretaceous tropical and subtropical central Pacific Ocean. This volume.

Thiede, J. & van Andel, T. H. 1977. The paleoenvironment of anaerobic sediments in the late Mesozoic South Atlantic Ocean. *Earth and Planetary Science Letters* **33**, 301–309.

Thierstein, H. R. 1979. Paleooceanographic implications of organic carbon and carbonate distribution in Mesozoic deepsea sediments. *Deep Drilling Results in the Atlantic Ocean: Continental Margins and Paleoenvironment* (Eds M. Talwani, W. W. Hay & W. B. F. Ryan), Washington: American Geophysical Union, Maurice Ewing Series, Vol. 3, pp. 249–274.

Tissot, B., Demaison, G., Masson, P., Delteil, J. R. & Combaz, A. 1980. Paleoenvironment and petroleum potential of middle Cretaceous black shales in Atlantic basins. *American Association of Petroleum Geologists Bulletin* **64**, 2051–2063.

Tissot, B., Deroo, G. & Herbin, J. P. 1979. Organic matter in Cretaceous sediments of the North Atlantic: contributions to sedimentology and paleogeography. *Deep Drilling Results in the Atlantic Ocean: Continental Margins and Paleoenvironment* (Eds M. Talwani, W. W. Hay & W. B. F. Ryan), Washington: American Geophysical Union, Maurice Ewing Series, Vol. 3, pp. 362–374.

Tourtelot, H. A. 1964. Minor-element composition and organic carbon content of marine and nonmarine shales of Late Cretaceous age in the western interior of the United States. *Geochim. et Cosmochim. Acta* **28**, 1579–1604.

Tourtelot, H. A. 1979. Black shale—its deposition and diagenesis. *Clays and Clay Minerals* **27**, 313–321.

Tucholke, B. E. & Vogt, P. R. 1979. Western North Atlantic: Sedimentary evolution and aspects of tectonic history. *Initial Reports of the Deep Drilling Project* **43**, 791–825.

Vine, J. D. & Tourtelot, E. B. 1969. Geochemical investigations of some black shales and associated rocks. *U.S. Geological Survey Bulletin 1314-A*, 43pp.

Vine, J. D. & Tourtelot, E. B. 1970. Geochemistry of black shale deposits—a summary report. *Economic Geology* **65**, 253–272.

Vinogradov, A. P. 1953. The elementary chemical composition of marine organisms. *Sears Foundation for Marine Research Memoir 2*, 647pp.

Volkov, I. I. & Fomina, L. S. 1974. Influences of organic materials and processes of sulfide formation on distribution of some trace elements in deep-water sediments of Black Sea. *American Association of Petroleum Geologists Memoir* **20**, 457–476.

Wedepohl, K. H. 1964. Untersuchung am Kupferschiefer in Nordwest deutchland; ein Beitrag zur Deutung der Genese bituminoger Sediment. *Geochim. et Cosmochim. Acta* **28**, 305–364.

4. Oxygen-deficient Depositional Paleoenvironments in the Mid-Cretaceous Tropical and Subtropical Central Pacific Ocean

J. Thiede

Department of Geology, University of Oslo

W. E. Dean and G. E. Claypool

U.S. Geological Survey, Denver

Complete sections of organic carbon-rich Aptian–Cenomanian limestones have recently been sampled in the central North Pacific Ocean. They are contained in a sequence of pelagic strata deposited in oxygen-deficient environments which had developed along the flanks of the Mid-Pacific Mountains and Hess Rise, two aseismic rises which since have subsided 1–3 km. The "anoxic" strata in the Mid-Pacific Mountains consist of beds of dark, organic carbon-rich, laminated limestone of Early Aptian age. These beds of organic carbon-rich limestone are interbedded with volcanic ash and were deposited when the Mid-Pacific Mountains were still situated about 20° south of the Upper Mesozoic paleoequator. Beds containing high concentrations of organic carbon are only 3–70 cm thick and represent only several tens of thousands of years each. Apparently the depositional environment was very unstable during this time because the "anoxic" intervals are of short duration. Organic carbon-rich, laminated limestones on Hess Rise are of Late Aptian to Early Cenomanian age, a time period when this aseismic rise was passing under the equatorial divergence with its high surface water productivity. These two occurrences of "anoxic" strata can be used to develop models which apply also to other mid-Cretaceous central Pacific Ocean "anoxic" deposits on Manihiki Plateau and on Shatsky Rise. Pyrolysis assay results indicate that most of the organic matter in the limestones on Hess Rise is composed of lipid-rich kerogen derived from aquatic marine organisms. The limestones from the Mid-Pacific Mountains tend to have low ratios of pyrolytic hydrocarbons to organic carbon and low hydrogen indices, suggesting that the organic matter contains some land-derived, humic-rich material, which probably came from numerous volcanic islands that must have existed when the site was considerably shallower, in addition to material of marine origin. The organic carbon-rich strata were deposited during times of high pelagic sedimentation rates in the mid-Cretaceous central Pacific Ocean which coincide with high accumulation rates of pelagic biogenic sediment components in the Indian, Atlantic and Pacific Oceans, and which therefore suggest that the mid-Cretaceous was a period of high productivity of the oceanic surface water masses. However, oxygen-deficiency developed in the central Pacific Ocean only when certain structural and hydrographic factors coincided during a time of relatively sluggish bottom water circulation.

1. Introduction

Marine depositional environments that resulted in the accumulation of "anoxic" sediments overlain by oxygen-deficient water masses have developed throughout the Phanerozoic. These sediments can contain relatively high concentrations of organic matter and they are considered very important source rocks for hydrocarbons. The temporal distribution of hydrocarbons suggests (Irving *et al.*, 1974) that most of the known oil formed during Late Mesozoic times, especially during the mid-Cretaceous. Such environments have been deduced from sediments deposited in many ancient shelf seas which covered stable cratonic areas (Hallam, 1967; Kemper & Zimmerle, 1978), and from the folded rocks of old tectonic mobile belts (Jenkyns, 1980). Their formation was believed by many to be independent of ocean-wide mechanisms (Kemper & Zimmerle, 1978), and it was surprising when black or very dark deposits of varying lithologies but all rich in organic matter, were also found in the pelagic realm in the course of deep-sea drilling during the past twelve years. These oceanic strata are also mostly of Cretaceous age and occur in all major ocean basins. It has been possible to collect long and relatively detailed sections of organic carbon-rich strata from the Atlantic and Indian Oceans during early stages of deep-sea drilling, but the Pacific record consisted only of a few scattered, in part dubious, samples.

Only recently, during Leg 62 of the Deep-Sea Drilling Project (DSDP), were complete records of "anoxic" strata recovered at continuously-cored drill sites on the flanks of the Mid-Pacific Mountains and Hess Rise in the central North Pacific Ocean (Dean *et al.*, 1981). The new data allow constraints to be placed on the boundary conditions for oxygen-deficient pelagic depositional environments. In this paper we therefore want to review the new data from the Mid-Pacific Mountains and from Hess Rise, and to compare the new information with data from previously known occurrences (Schlanger & Jenkyns, 1976) from the Pacific Ocean (Figure 1, Table 1). To interpret the oxygen-deficient pelagic paleoenvironments it will be

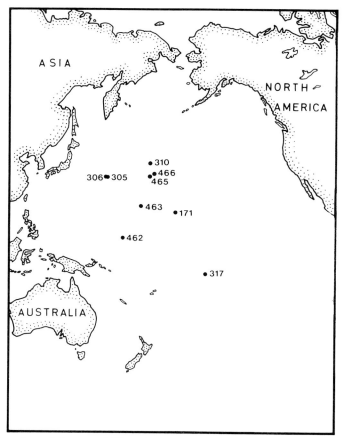

Figure 1.   Locations of deep-sea drilling sites in the Pacific Ocean where mid-Cretaceous organic carbon-rich volcanic or pelagic deposits have been recovered (cf. Table 1).

Table 1. Location, present water depth, lithology, age and estimated paleodepth of deposition of mid-Cretaceous organic carbon-rich deposits in the tropical and subtropical central Pacific Ocean. The information has been compiled from data published in the site reports of the Initial Report volumes of the Deep-Sea Drilling Project (see text for references).

Location	Latitude	Longitude	Water depth	Lithology of organic carbon-rich strata	Age of organic carbon-rich strata (m.y. B.P.)	Estimated depth of deposition (m)
Site 171 (Leg 17, Mid-Pacific Mountains, Horizon Guyot)[1]	19°07.9'N	169°27.6'W	2295	Laminated calcareous volcanic sandstone	Turonian (86–88)	400–800
Site 305 (Leg 32, Shatsky Rise)[2]	32°00.1'N	157°51.0'E	2903	Carbonaceous zeolitic pelagic shale	Cenomanian (98–100)	1000–2000
Site 306 (Leg 32, Shatsky Rise)	31°52.0'N	157°28.7'E	3399	Carbonaceous silicified radiolarian shale	Aptian–Barremian (~115)	1000–2000
Site 310 (Leg 32, Hess Rise)[3]	36°52.1'N	176°54.1'E	3516	Laminated carbonaceous pelagic shale	Albian–Cenomanian (~100)	1000–2000
Site 317 (Leg 33, Manihiki Plateau)[4]	11°00.1'S	162°15.8'W	2625	Volcanic sandstone	>Barremian (>120)	300–1000
Site 463 (Leg 62, Mid-Pacific Mountains)	21°21.0'N	174°40.1'E	2525	Laminated silicified limestone with volcanic ash	Early Aptian (112–113)	500–1500
Site 465 (Leg 62, Hess Rise)	33°49.2'N	178°55.1'E	2161	Laminated limestone with volcanic ash	Late Aptian–Early Cenomanian (98–103)	500
Site 466 (Leg 62, Hess Rise)	34°11.5'N	179°15.3'E	2665	Laminated limestone with volcanic ash	Late Aptian–Early Cenomanian (98–103)	500–1000

[1] Based on 2 samples (2.8% and 2.3% C_{org}).
[2] One horizon only.
[3] One fragment in core catcher (no C_{org} data).
[4] One sample (28.7% C_{org}).

necessary to describe the composition, texture and sedimentary structures of the "anoxic" deposits and to try to reconstruct when, and at what depth and latitude, they were deposited.

The detection of "anoxic" strata in the deep-sea initiated immediately an intense discussion about the factors which might have led to the oxygen deficiency at the ocean bottom (Degens & Stoffers, 1976). Observations from the Mediterranean and South Atlantic seemed to support an "euxinic" model (Ryan & Cita, 1977), whereas the Pacific samples seemed to suggest an "oxygen minimum" model (Schlanger & Jenkyns, 1976). The variable composition of the rocks containing high concentrations of organic matter (Thiede & van Andel, 1977), the differences in composition of the organic matter in the rocks (Tissot *et al.*, 1979), and the wide spread in time (almost 40 m.y.) during the Cretaceous period when these rocks were deposited (Arthur & Natland, 1979) have as yet defied a unifying explanation.

We will concentrate on Pacific Mesozoic open pelagic deposits. Cenozoic laminated organic carbon-rich sediments which have recently been drilled in the Gulf of California (Curray, Moore *et al.*, 1979) accumulated in a restricted, young, immature basin (Thiede, 1978) rather than in an old, open and mature ocean basin such as the Late Mesozoic Pacific Ocean. Quaternary black organic carbon-rich hemipelagic muds are presently accumulating along the Peruvian continental margin, but they are believed to be due to the development on an oxygen-deficient depositional environment in a region where the oxygen minimum under the coastal upwelling zone impinges on the outer shelf and upper continental slope. So far, none of the modern analogs is applicable to explain the development of oxygen-deficient paleoenvironments in the open Late Mesozoic central Pacific Ocean.

2.   Occurrences of mid-Cretaceous organic carbon-rich "anoxic" sediments in the Pacific Ocean

Most of the descriptions and analytical data discussed here are from Sites 463, 465 and 466 which were drilled in the central North Pacific Ocean during DSDP Leg 62 (Dean *et al.*, 1981). All three sites (Figure 1) are situated on structural highs (Mid-Pacific Mountains and Hess Rise) which rise 2–3 km above the surrounding 5–6 km-deep basin floor and which were at or near sea-level during the Early Cretaceous (Hamilton, 1956). Thin beds of organic carbon-rich "anoxic" strata (Table 1) were recorded at Site 171 on Horizon Guyot in the eastern Mid-Pacific Mountains on DSDP Leg 17 and on Hess Rise during DSDP Leg 32 Winterer, Ewing *et al.*, 1973; Larson, Moberly *et al.*, 1975a. Organic carbon-rich strata have also been recovered from other aseismic rises in the central Pacific Ocean, namely Shatsky Rise (DSDP Leg 32, Sites 305 and 306, Larson, Moberly *et al.*, 1975b and c) and on Manihiki Plateau (DSDP Leg 33, Site 317, Schlanger, Jackson *et al.*, 1976).

Our discussion will be restricted to laminated sediments which lack an autochthonous, benthic macrofossil fauna, which have reduced microfossil assemblages, e.g. foraminifers (Boersma, 1981), and which do not contain any traces of burrowing organisms. Such strata probably reflect truly anaerobic bottom water conditions (Byers, 1977).

2.1.   Mid-Pacific Mountains

Some of the best Pacific sections through mid-Cretaceous "anoxic" strata have been obtained from drill sites at the flanks of the Mid-Pacific Mountains. The earliest record of such sediments was found on Horizon Guyot (DSDP Leg 17, Site 171) at the eastern end of the Mid-Pacific Mountains, where greenish-gray, partly slightly mottled, partly finely laminated calcareous volcanic siltstone and claystone of Turonian age containing more than 2% of organic carbon (17-171-25) were recorded. They overlie a sequence of volcanic sandstone, shallow water limestone and basalt of probable Cenomanian age.

Three beds of siliceous limestone containing more than 2% organic carbon occur within a Lower Aptian pelagic limestone sequence (Figure 2) at DSDP Site 463 in the western Mid-Pacific Mountains. The lithologic unit rich in organic matter overlies pelagic limestones interbedded with layers of displaced shallow-water-derived carbonates whose thickness, grainsize and frequency increases downwards, and underlies a sequence of Upper Mesozoic and Cenozoic calcareous pelagic sediments (Thiede, Vallier et al., 1981). The organic carbon-rich intervals at Site 463 belong to the Lower Aptian *Chiastozygus litterarius* nannofossil zone (Cepek, 1981). The age scale in Figure 2 is based on a linear interpolation between the upper and lower boundaries of this zone which lasted from approximately 112 to 115 m.y. B.P. (van Hinte, 1976). Assuming a constant sedimentation rate across this time span the ages of the individual intervals with high concentrations of organic carbon and their duration can be estimated to within 10 000 years.

The earliest and shortest of the three laminated and organic carbon-rich intervals began about 113.2 m.y. B.P., the second one about 113.18–113.19 m.y. B.P., and the third one about 113.08–113.09 m.y. B.P. The occurrence of laminated intervals within dark gray, mottled, silicified limestone suggests that a depositional environment may have developed which was favorable for the preservation of organic

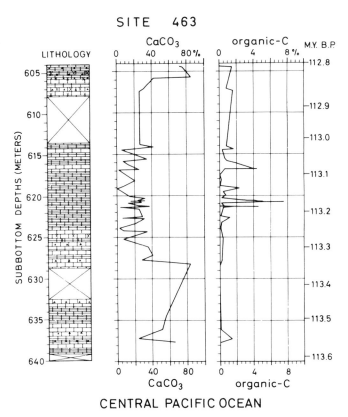

SITE 463

Figure 2. Distribution of calcareous and carbonaceous matter in the Lower Aptian tuffaceous and silicified limestone of Site 463 (DSDP Leg 62, Mid-Pacific Mountains).

Figure 3.   Sedimentary structures of Lower Aptian tuffaceous and carbonaceous limestone of DSDP
    Site 463, Leg 62, Mid-Pacific Mountains). (A) 62-463-70-5 (cm 80–110), (B) 62-463-71-1 (cm
    115–140), (C) 62-463-67-2 (cm 70–90).

carbon. Many of the laminated intervals, however, do not contain particularly high
concentrations of organic carbon. The three organic carbon-rich intervals of this site
all have very sharp lower boundaries indicating that oxygen deficiency was
established rapidly. The upper boundaries are gradational, suggesting that it took a
few thousand to a few tens of thousands of years for the environment to return to its
original oxygenated condition.

The rocks in Cores 70 and 71 in Hole 463 (623.0–632.5 m) are dominantly highly silicified limestones with some chert and a few layers of volcanic ash. Radiolarians are abundant but they are poorly preserved (recrystallized and partly dissolved, cf. Figure 4), compared with the much better preserved radiolarians above and below. This observation suggests that the source of silica for diagenetic silicification of the limestones was biogenic, and hence there was a high surface-water fertility although the site was situated far south of the equatorial divergence during Early Cretaceous time (Sayre, 1981). As a result of silicification the concentration of $CaCO_3$ is generally less than 30% and the concentration of Si are as high as 40% (Dean, 1981).

The most remarkable sedimentary structures in Cores 69 to 72 are the laminations in the dark gray, clayey intervals that are intercalated with the bluish-white to dark greenish-gray, mottled limestones. The laminated intervals range from a few centimeters to more than 60 cm in thickness, but only three of them contain concentrations of organic carbon greater than 2%. The darker intervals occur throughout Cores 69 (604.0–613.5 m) to 72 (632.5–642.0 m), but they are most abundant in Cores 70 and 71 where it is difficult to distinguish them from thin volcanic ash layers.

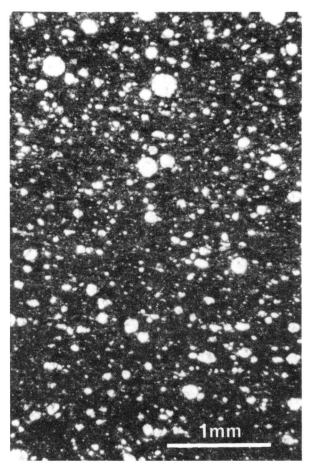

Figure 4.   Recrystallized radiolarians in organic carbon-rich siliceous limestone of Site 463 (Mid-Pacific Mountains). 62-463-70-5 (cm 134–137). Cross polarized light. Bar = 1 mm.

2.2. Shatsky Rise

Sites 305 (2903 m water depths) and 306 (3399 m water depths) where thin horizons of "anoxic" sediments have been penetrated, are both situated on the southern flank of Shatsky Rise in the north-western Pacific Ocean (Larson, Moberly *et al.*, 1975*b* and *c*). The organic carbon-rich deposits at Site 305 (32-305-37) consist of a three-centimeter-thick layer of zeolitic pelagic shale of Early Turonian to Early Cenomanian age which is intercalated into a nannofossil chalk sequence with many chert horizons. One sample of this shale contained 9.3% organic carbon. Thin layers of silicified radiolarian shale which are intercalated into a chert and porcellanite and therefore poorly cored and recovered chalk sequence of Aptian to Barremian age (32-306-12, 32-306-12), represent the "anoxic" sediments found at Site 306. It is noteworthy that these occurrences are very thin (approx. 1 cm thick) and that they have occurred over a time span of 15–20 m.y. during the mid-Cretaceous. Numerous burrows in the cherts above and below the "anoxic" horizons, which have sharp boundaries, document that the "anoxic" sediments reflect very short-lived, very sudden events of probably limited regional extent.

2.3. Manihiki Plateau

Site 317 is situated in approximately 2600 m water depths on a part of the Manihiki Plateau south-west of Manihiki Island. One sample (Core 16A-2-133) of volcanic sandstone contained 28.7% organic carbon (Schlanger, Jackson *et al.*, 1976). This sample, in addition to other observations from Shatsky and Hess Rise, has given rise to a hypothesis that oceanic midwater oxygen minima during mid-Cretaceous time resulted in the development of ocean-wide oxygen-deficient depositional environments (Schlanger & Jenkyns, 1976; Jenkyns & Schlanger, 1981). Reddish-brown horizons in the sediments just above and below the organic carbon-rich sample document that the reducing environment cannot have persisted very long, provided the reddish hues are not due to alterations by hydrothermal solutions. The underlying basalts which are approximately 106–107 m.y. old (Jackson & Schlanger, 1976) represent the surface of a thick section of oceanic ridge tholeiite type basalt which extruded since approx. 110 m.y. B.P. to construct this rise. The big vesicles suggest that the eruptions happened in shallow water depths (less than 400 m), which is also supported by the occurrence of large molluscs in the overlying volcanoclastic sedimentary sequence.

2.4. Hess Rise

A small piece of black pelagic shale of Early Cenomanian to Late Albian age was recovered from Site 310 (32-310-17) on the central part of Hess Rise. This shale must be very thin, because it is represented by one small fragment in the core catcher. The shale bed is intercalated into a chert-rich sequence of nannofossil chalks, porcellanites and zeolitic pelagic clays. The sample was not analyzed for organic carbon, but it had a bituminous odor and burned when heated. The recovery of the mid-Cretaceous sediments at this site has been very poor because of the occurrence of frequent chert layers that it remained unknown if further horizons of "anoxic" strata have been penetrated.

The organic carbon-rich deposits which were recovered at Sites 465 (Figure 5) and 466 on southern Hess Rise are different from previous occurrences mainly because they are several tens of meters thick, they are organic carbon-rich throughout, they are laminated but they also contain sedimentary structures indicative of erosive bottom water currents.

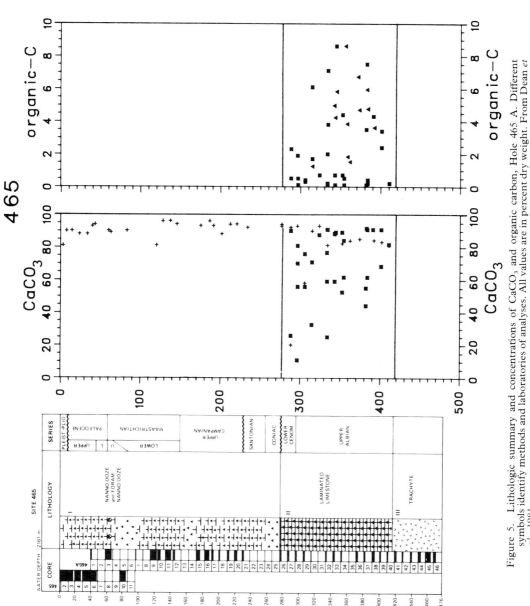

Figure 5. Lithologic summary and concentrations of CaCO₃ and organic carbon, Hole 465 A. Different symbols identify methods and laboratories of analyses. All values are in percent dry weight. From Dean *et al.*, 1981.

Figure 6.   Sedimentary structures in organic carbon-rich limestone of Site 465 on southern Hess Rise.
(A) 62-465-40-1 (cm 90–110), (B) 62-465A-36-3CC (cm 20–30).

The organic carbon-rich laminated limestone at Site 465 (Figures 5 and 6) on southern Hess Rise is 136 m thick and spans about 5 m.y. from Late Albian to Early Cenomanian. This limestone overlies altered trachytic volcanic basement which formed by subaerial differentiated off-ridge oceanic island volcanism (Seifert *et al.*, 1981). The lateral equivalent of this laminated limestone at Site 466 consists of faintly-laminated nannofossil chalk and limestone that are similar to the laminated limestone at Site 465 but slightly lighter in color. However, the organic carbon-rich chalk and limestone of Site 466 span only about one million years during the Late Albian in about 66.5 m of the section.

The laminated appearance of the limestones at Sites 465 and 466 is the result of concentrations of radiolarians and foraminifers and of flaser-like laminae containing streaks of dark, organic (?) material in a matrix of micritic calcite (Figure 7). Chert is common throughout the laminated limestone unit, but silicification of the limestone is slight and confined mainly to rare, thin interbeds of gray, massive to faintly-laminated limestone. Radiolarians are common but most have been replaced by calcite. Varying degrees of silicification have resulted in considerable variation in concentrations of carbonate in the laminated limestones although most samples contain more than 60% $CaCO_3$. The laminated limestones also contain common redeposited fossils, including mollusc fragments, shelf benthic foraminifers and radiolarians (Boersma, 1981).

Concentrations of organic carbon in samples of the laminated limestone from Site 465 are as high as 8.6%, but most are less than 5% (Figure 5). There are no apparent differences between the laminated limestones containing higher concentrations of organic carbon and those containing lower concentrations. The organic carbon contents in the correlative laminated chalk and limestone unit at Site 466 is similar to that in the laminated limestone at Site 465.

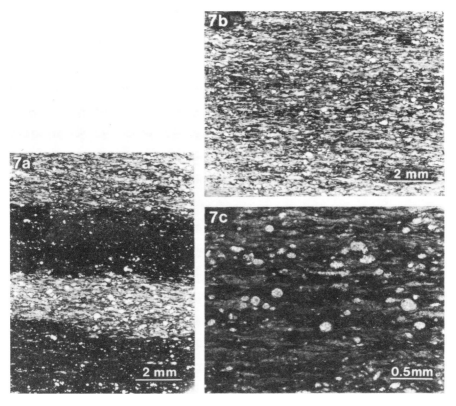

Figure 7. Micrographs of laminated organic carbon-rich limestone of Hole 465A, showing concentrations of radiolarians and streaks of dark organic matter in a matrix of micritic calcite. (a) 62-465A-28-2 (cm 53). Laminated chert layer: dark layer is completely silicified olive-gray limestone, light layer is only partly silicified. Cross-polarized light, bar = 2 mm. (b) Detail of (a). Typical example for the laminated limestone of this site. Radiolarians and foraminifers have been partly to completely replaced and filled with silica in a micritic carbonate matrix. The laminated appearance of the rock is due to wispy, discontinuous, flaser-like laminae of dark, organic (?) material and some sorting of radiolarians and foraminifers into layers. Plane light, bar = 2 mm. (c) Detail of (b).

3. Temporal distribution of mid-Cretaceous "anoxic" sediments

Many of the sections containing rocks deposited in an oxygen-deficient environment in the central Pacific during mid-Cretaceous time are difficult to date because of the relatively poor knowledge of the Middle and Lower Cretaceous microfossil assemblages and because of the poor correlation between Upper Mesozoic biostratigraphy and magnetostratigraphy. Dissolution and diagenetic changes typical of most Upper Mesozoic pelagic rocks in the Pacific Ocean have further reduced the use of biostratigraphy to determine in detail the timing of the development of the oxygen-deficient depositional paleoenvironments.

The ages indicated in Table 1 are based on the biostratigraphies listed in the site reports and their correlation with the Cretaceous time scale of van Hinte (1976). The ages of pelagic strata deposited under oxygen-deficient conditions in the Late Mesozoic Pacific Ocean are spread over 30–35 m.y. of the Cretaceous period from Barremian to Turonian. Most of the layers of "anoxic" strata are relatively thin and were deposited during events of a few thousand years in duration (as exemplified by the data from Site 463) whereas Sites 465 and 466 on southern Hess Rise document the existence of bottom-water oxygen deficiency for a time span of a few million years. Rocks recovered from the Mid-Pacific Mountains and Shatsky Rise drill sites suggested repeated development of oxygen-deficient bottom water but the individual events could only be observed in one drill site. We cannot determine if this is due to the limited regional extent of the oxygen-deficient bottom water or if this is an artifact of poor core recovery.

4. Depth of deposition

All locations where oxygen-deficient bottom water conditions developed in the mid-Cretaceous central Pacific Ocean are situated on the flanks of aseismic structural highs of volcanic origin. The main platforms of these highs rise 2–4 km above the surrounding very deep Mesozoic ocean floor (Pitman *et al.*, 1974). Because these aseismic rises were always above the calcite compensation depth (CCD) they are capped by thick calcareous Mesozoic and Cenozoic calcareous deposits. Few sites on these rises have penetrated volcanic basement and therefore the amount of subsidence which occurred after the volcanic activity that formed the aseismic rises cannot be determined. Autochthonous as well as displaced calcareous benthic fossils which lived in shallow waters close to the sea surface have been found in the deepest parts of the sections recovered at many drill sites (Thiede *et al.*, 1981) or have been dredged from the flanks of many seamounts (Hamilton, 1965; Matthews *et al.*, 1974) in the central subtropical Pacific Ocean. These shallow-water fossils suggest that the volcanic edifices under the calcareous strata were once close to, and in many instances even above, the sea surface. It is often difficult to establish the rates of subsidence of aseismic rises (Schlanger *et al.*, 1981) because it is not clear if they formed by volcanism very close to the Upper Mesozoic spreading center in the central Pacific, or if they were built by differentiated off-ridge volcanic activity. In the first case one can assume that they followed the thermal cooling curve of the surrounding oceanic basaltic crust (Sclater *et al.*, 1971), although at a shallower level. In the second case, the steep part of the thermal cooling curve during the phase of rapid subsidence of the young crust would have been truncated. Studies of the volcanic basement rocks of Hess Rise (Vallier *et al.*, 1981) suggest differentiated subaerial volcanism which is typical for oceanic volcanic islands.

The incomplete data available from the eight drill sites with "anoxic" sediments in the central Pacific Ocean (Table 1) make any attempt to reconstruct quantitatively

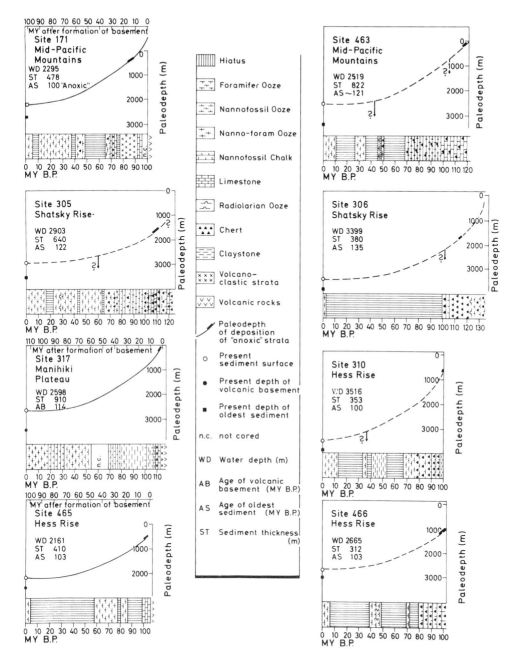

Figure 8. Reconstruction of the depth of deposition of central Pacific deep-sea drilling sites where organic carbon-rich mid-Cretaceous volcanic and pelagic strata have been observed. The construction of the paleodepth of deposition of several sites is quite tentative because essential data are missing. Paleontological evidence from Sites 465 and 466 (Boersma, 1981) suggests that Hess Rise subsided considerably slower than young crust close to oceanic spreading centers.

rates and amounts of subsidence highly tentative and speculative (Figure 8), but it is often possible to establish minimum values. Independent paleontological evidence (Boersma, 1981) has been used to adjust and to verify the reconstructions. These reconstructions indicate that the oxygen-deficient depositional environments developed in estimated depths ranging from 300 to maximal 2000 m (Table 1, Figure 8). These environments represent relatively shallow depths where intermediate water masses abutted the elevated sea floor of structural highs.

5. Paleogeography

It is clear from the composition of the volcanic rocks (Vallier *et al.*, 1981), from the paleontologic data and from sedimentologic evidence, that the summits of most of the aseismic rises in the central Pacific once reached into very shallow waters. Indications of shallow-water depositional environments have also been found considerably below their summits. It can be assumed that the dominant portion of the volcanic edifices whose flanks are now hidden below thick carbonate deposits, has been generated prior to their final subsidence. Many of these edifices must have supported large volcanic islands or groups of islands similar to the modern volcanic

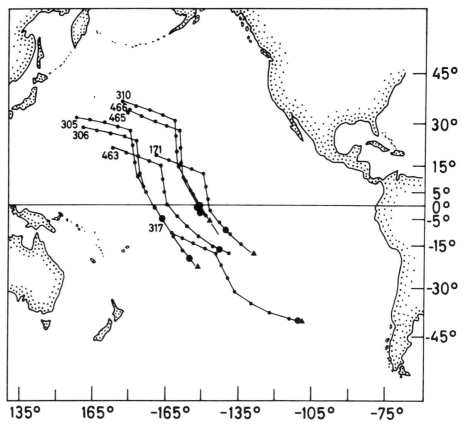

Figure 9.   Path of plate rotations of the deep-sea drill sites in the central Pacific (after Lancelot, 1978) where mid-Cretaceous paleoenvironments favored the preservation of high organic carbon concentrations (cf. Table 1). Location of sites during deposition of organic carbon-rich sediments is marked by a large dot.

oceanic islands in the Pacific Ocean with summits that may have been up to several kilometers above sea level. These islands must have been important obstacles for the circulation of the oceanic surface water.

In addition to subsidence from close to sea-level in the Late Mesozoic to 1–3 km water depth at present, the aseismic rises have also undergone substantial horizontal movement as a result of movement of the Pacific crustal plate toward the north-west (Lancelot, 1978). All eight sites underlain by "anoxic" strata (Figure 9) are located on aseismic rises which formed south of the equator. Except for Manihiki Plateau these sites are now in the central North Pacific Ocean. By reconstructing the paleolatitudes of sites underlain by Upper Mesozoic "anoxic" strata using the model of Lancelot (1978), it is evident that with the exception of the sites on Hess Rise, all sites were south of the equator during the mid-Cretaceous, the Hess Rise sites were about at the equator (Figure 9).

6. Composition of the carbonaceous sediments

The lithologies of organic carbon-rich rocks from the Upper Mesozoic central Pacific Ocean are highly variable, despite the fact that they were all deposited in true pelagic environments on aseismic rises. Because these sites are elevated above the surrounding ocean basin floor, they have been protected against massive fluxes of suspended terrigenous matter of continental origin although they can have received minor quantities of airborne material. A discussion of the composition of the organic carbon-rich rocks therefore can be subdivided into organic carbon contents, other biogenic components and inorganic components.

6.1. The organic carbon contents

The organic carbon concentrations of most mid-Cretaceous "anoxic" lithofacies is greater than 2% and may be as much as 20–30% of the bulk sediment (Schlanger, Jackson et al., 1976). The origin and composition of this organic material can be characterized by various methods, such as petrographic description of discernible particles, stable carbon isotope measurement, pyrolysis assay, detailed gas chromatographic analyses, to name but a few.

The observation of carbonized vascular plant debris in Lower Aptian sediments at Site 463 (Timofeev & Bogolyubova, 1981; Mélières et al., 1981) and in an 80-meter-thick horizon of Upper Aptian gray claystones at Site 462 in the Nauru Basin (Jenkyns & Schlanger, 1981) suggests that there may have been a considerable influx of terrigenous organic matter from nearby land areas. These observations are supported by pyrolysis analyses (Dean et al., 1981) of Site 463 samples which pointed to a relatively low hydrogen index, but high oxygen index, and whose organic matter (Figure 10) appears to contain an important fraction of Type III kerogen (Tissot et al., 1974). Some of the samples of Site 463, however, have a hydrogen index that is intermediate between Type II and Type III kerogen (Figure 10) suggesting that the organic carbon-rich horizons may contain variable mixtures of Type II (typical marine) and Type III (terrestrial or highly oxidized) matter. The proportion of marine organic material generally increases with increasing organic carbon concentrations. The same samples also contain high molecular weight, odd-numbered n-alkanes which are derived from cuticular waxes of higher plants (Dean et al., 1981), but which indicate that the contribution of terrestrial organic matter was probably minor. Mélières et al. (1981) have demonstrated that several episodes of deposition of organic matter can be discerned in the mid-Cretaceous section of Site 463. They are composed of a lower interval of light

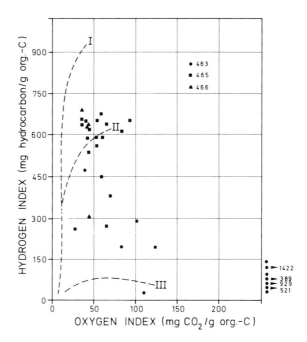

Figure 10.   Scatter plot of hydrogen and oxygen indices for Holes 463, 465A and 466 (from Dean *et al.*, 1981). Roman numerals refer to thermochemical transformation pathways of three different types of sedimentary organic matter (Tissot *et al.*, 1974): (I) Algae-rich or microbially reworked organic matter; (II) Marine organic matter; (III) Organic matter from terrestrial plants and highly weathered, chemically inert organic matter of unknown origin.

colored, burrowed sediments with undifferentiated organic matter, an intermediate interval with organic matter of terrestrial and marine origin (either mixed or alternating), and an upper interval of very dark, laminated sediments with organic matter entirely of marine origin.

The organic matter found in the organic carbon-rich laminated Upper Albian and Lower Cenomanian limestones on Hess Rise has a composition different from that at Site 463. Pyrolysis data indicate that most of the samples from Hess Rise contain Type II kerogen whereas only a few samples have organic matter that is intermediate between Type II and III kerogen (Figure 10). Based on lipid analyses in the samples of mid-Cretaceous limestone from Sites 465 and 466, Simoneit (1981) suggested that the organic matter is dominantly of autochthonous marine origin whereas presence of elemental sulfur suggested an euxinic depositional environment. There are no indications of contributions from vascular plant residues to the lipids, but there is evidence for bacterial lipid residues.

6.2.   Other biogenic components

All pelagic sections (except the volcanic rocks, see Table 1) which contain intervals of "anoxic" organic carbon-rich strata originally contained calcareous and siliceous components which now appear as alterations of chalk or limestone and chert layers. These strata represent depositional environments of relatively high rates of sedimentation of biogenic pelagic components. Clinoptilolite, quartz (chalcedony) opal CT (Mélières *et al.*, 1981) are probably diagenetic products related to siliceous fossils. Barite which occurs in minor amounts may also be biogenic (Dean &

Schreiber, 1978) because it is correlated to a silica-rich depositional environment (Mélières *et al.*, 1981). High concentrations of pyrite, present in these sediments, are taken to correlate with concentrations of high organic carbon.

Shallow-water derived fossils have been observed in the organic carbon-rich intervals at Sites 465 and 466 (Boersma, 1981) as well as in the underlying sediment section at Site 463. They point to the former existence of adjacent neritic environments which because of their locations in the tropical and subtropical central Pacific Ocean (Figure 9) led to the build-up of thick and extensive carbonate deposits.

6.3.   The non-organic sediment fraction

Although the components in "anoxic" strata are dominantly biogenic, they also contain a large fraction of non-biogenic material. There is a correlation between the laminated organic carbon-rich deposits and volcanogenic deposits (Table 1) which consist of volcanic sandstones, ash layers or clay horizons whose clay minerals have been derived mostly from altered volcanic products. The dominant portion of the clay minerals at Site 463 (Mélières *et al.*, 1981) consist of smectite (10 Å phyllosilicate structures) and illite which probably evolved diagenetically from smectite. This composition is in contrast to the "black shales" in the Atlantic Ocean (Chamley & Robert, this volume) where well-crystallized smectites are the most abundant minerals in the Late Mesozoic organic carbon-rich rocks and in the under- and overlying organic carbon-poor rocks. The Atlantic smectites are probably derived from a terrigenous source on the adjacent continents which experienced a warm climate with strong contrasts in seasonal humidity.

The Site 463 sediments also contain a considerable amount of amorphous material which consists mainly of volcanogenic alumino-silicates (Mélières *et al.*, 1981); feldspars (both potash feldspar and plagioclase) have been found throughout the "anoxic" sediments at Site 463 and are probably directly or indirectly of volcanogenic origin. Rea & Harrsch (1981) and Rea & Janecek (1981) have also detected minor amounts of airborne material in the sediments at Site 463 and Sites 465 and 466 on Hess Rise.

7.   Discussion and models for the development of oxygen-deficient depositional paleoenvironments in the Upper Mesozoic central Pacific Ocean

The descriptive results of the occurrences of Upper Mesozoic "anoxic" strata in the Pacific Ocean are not easily reconciled with most previous hypotheses which have been applied to explain the evolution of oxygen-deficient paleoenvironments in the open pelagic realm (euxinic and oxygen minimum models). The small number of Pacific drill sites (Table 1) where such sediments have been found, prevents a thorough discussion, but the data obtained from these drill sites allow us to define some boundary conditions controlling the development of oxygen-deficient paleoenvironments in the Upper Mesozoic central Pacific Ocean and to define a number of models to explain these occurrences of "anoxic" strata. These models will have to satisfy the following diverse observations.

(1)   There are two types of oxygen-deficient depositional environments. Both developed in intermediate water depths along the flanks of large aseismic rises in the central tropical and subtropical Late Mesozoic Pacific Ocean. One of the oxygen-deficient environments led to very short-lived events of complete oxygen depletion,

probably in connection with volcanic activity, and is documented in the sediments by the coincidence of volcanogenic sediments and thin horizons of laminated organic carbon-rich rocks. These environments seem to be restricted to relatively small areas on the rises because they have not been sampled in adjacent drill sites. However, the poor recovery record of many Pacific Mesozoic sections, mainly because of abundant chert, limits the validity of this observation. Such environments may have repeatedly developed along the same rise in response to pulses of volcanic activity. All occurrences except those at Sites 465 and 466 on southern Hess Rise belong to this type which we will call the *organo-volcanic spike* model. The main characteristics of this depositional environment are its small regional extent, its short duration, its thin deposits and its coincidence with volcanic activity.

The other type of oxygen-deficient depositional paleoenvironment lasted several million years and led to the deposition of thick sequences of organic carbon-rich pelagic strata—which accumulated when the site crossed the region under the equatorial divergence as the Pacific crustal plate moved northward during the Mesozoic (Lancelot, 1978). This depositional environment which we will call the *equatorial crossing* model is documented at Site 465 and Site 466 on southern Hess Rise.

(2)   Oxygen-deficient depositional paleoenvironments developed in the central Pacific Ocean over a time span of 30–35 m.y. during Cretaceous time (Table 1); they do not represent isochronous events which affected the entire basin. Sites containing organic carbon-rich strata of mid-Cretaceous age are regionally restricted to certain areas which were spread over 45° of paleolatitude (Figure 9). These observations are not easily reconciled with the hypothesis (Jenkyns & Schlanger, 1981; see also Jenkyns, 1980 and Schlanger & Jenkyns, 1976) of the development of an expanded oxygen-depleted intermediate water mass during the mid-Cretaceous. Modern analogs suggest (Gardner *et al.*, 1980) that such environments can also develop under fertile surface water masses of subpolar and polar seas.

(3)   The contemporaneous existence of oxygenated bottom water on the basin floor elsewhere in the Pacific, the restriction of anaerobic paleoenvironments to the slopes of a number of aseismic rises (Table 1) in intermediate paleowater depths, and the intercalation of "anoxic" deposits in oxygenated pelagic sequences suggest that the laminated "anoxic" organic carbon-rich deposits are products of very exceptional paleoenvironments, both in time and space.

(4)   The presence of high concentrations of organic carbon in deposits which accumulated rapidly and which are the product of fertile surface water masses resulted from a combination of tectonic and hydrographic factors. The "anoxic" deposits are found on the slopes of aseismic rises that are now situated under several kilometers of water, but that during Late Mesozoic time probably reached close to the sea surface.

(5)   When Hess Rise crossed under the equatorial divergence approximately 100 m.y. B.P. (Figure 9), its shallower parts were situated well above the CCD and within a midwater oxygen minimum which existed under the equatorial divergence. Benthic foraminiferal faunas in these strata suggest that they were deposited in an upper bathyal paleoenvironment. Although the major part of the organic carbon-rich deposits are laminated, sedimentary structures such as cross bedding (Figure 6), the sorted nature of the pelagic rocks (Figure 7), and the hiatuses which interrupted the continuity of the sedimentary record during Cretaceous time (Figure 5) suggest that the depositional environments were situated in a few hundred meters water depth in the deep parts of the equatorial current regime of the Upper Mesozoic Pacific Ocean. Thin beds of volcanic ash, intercalated into these thick limestone

sequences, indicate that there were volcanic events of modest scale after the major constructive phases of Hess Rise volcanism had ceased.

(6)   The occurrences of thin layers of organic carbon-rich, laminated layers on the central Pacific aseismic rises are closely related to volcanic events. These rises and the islands they once supported interrupted the flow of surface currents in the Upper Mesozoic central Pacific Ocean. North and south of the equatorial current regime surface currents were probably sluggish within the central subtropical gyres, but relatively fertile surface waters may have been generated close to the former islands and submarine shoals due to current-induced upwelling of subsurface waters around the islands (Moore *et al.*, 1973) and due to additional nutrient supply from the islands. Debris of land plants and organic matter of terrigenous origin (Mélières *et al.*, 1981), the probable subaerial volcanism on top of some of the aseismic rises, the altered clay mineral assemblages, and fossil assemblages displaced from neritic environments around the islands are evidence for this scenario. The oxygen-deficient depositional environments resulting from increased productivity probably developed only locally, and had only a short life span until the sea floor was able to revert to its oxygenated stage.

(7)   The organic carbon-rich laminated deposits in the Upper Mesozoic Pacific Ocean are intercalated into pelagic deposits that accumulated relatively rapidly and therefore probably protected the organic carbon contents against rapid destruction (Müller & Suess, 1979). The high sedimentation rates and the combination of calcareous and siliceous fossils in these deposits suggest that they formed under relatively fertile and productive surface water masses above and around the aseismic rises.

(8)   The organic material of the laminated sediment layers consists of components of marine and terrigenous origin. The minor terrigenous components (Dean *et al.*, 1981; Mélières *et al.*, 1981) are restricted to deposits interpreted to have formed according to the organo-volcanic spike model whereas deposits that formed according to the equatorial crossing model led to the sedimentation of organic carbon of exclusively marine origin (Simoneit, 1981). This is different from Upper Mesozoic sequences of the North Atlantic Ocean which have been drilled close to the continental margins, and which contain organic matter often dominated by components of terrigenous origin (Deroo *et al.*, 1978; Tissot *et al.*, 1979) and dependent upon the lithologies of the host rocks (de Graciansky *et al.*, 1979).

(9)   The non-biogenic sediment components of the Upper Mesozoic deposits on central Pacific aseismic rises are largely volcanic in origin either because they were related directly to former active volcanism or because they represent the erosional products of ancient volcanic islands which were washed into the adjacent ocean basins (Mélières *et al.*, 1981). Neritic deposits containing shallow-water fossils are found around many of the Mesozoic central Pacific aseismic rises (Hamilton, 1956); they suggest that the summit areas of these complex volcanic massifs extended into the photic zone and that some of them supported rather larger tropical islands which were richly vegetated and which shed vegetal debris into the surrounding ocean basins. Contributions of airborne mineral components of continental origin are quantitatively minor (Rea & Harrsch, 1981; Rea & Janecek, 1981). In contrast, the clay minerals of the Atlantic mid-Cretaceous "black shales" came mainly from continental tropical source areas (Chamley & Robert, this volume).

(10)   The development of oxygen-deficient paleoenvironments in the Pacific during Late Mesozoic time coincided with similar environments in other ocean basins that existed within the same general mid-Cretaceous time period (Arthur & Natland,

1979), although the deposits within and between ocean basins are not strictly synchronous. They occurred in a mid-Cretaceous time interval around 80–120 m.y. B.P., when the global climate was warm, when eustatic sea levels were high, when spreading rates were fast, when volcanic events affected the central Pacific Ocean and when pelagic sediments in the world ocean accumulated relatively fast (Fischer & Arthur, 1977; Worsley & Davies, 1979). Evidence deduced from frequency and age distribution of reworked and displaced pelagic sediment components suggests very sluggish bottom-water current regimes (Thiede *et al.*, 1981) in the mid-Cretaceous Pacific Ocean which in most instances were probably able to renew the oxygen contents of the oceanic bottom waters, but which were slow enough to allow the depletion of dissolved oxygen in areas of high fluxes of organic material to the sea floor, even in regions at a distance from the equatorial upwelling regimes. This mechanism could explain the contemporaneous existence of oxygenated deep water and oxygen-depleted intermediate water in the mid-Cretaceous central Pacific. It therefore does not seem necessary to postulate the development on an extensive mid-water oxygen minimum (Jenkyns, 1980), which would have to extend across 30–40° of latitudes at least, or of euxinic conditions (Simoneit, 1981) in an ocean basin. Each of the Pacific occurrences of mid-Cretaceous "anoxic" strata was preserved due to the coincidence of several tectonic and oceanographic factors. These coincidences may have occurred elsewhere and at different times in open and mature pelagic paleoenvironments such as in the upper Mesozoic Pacific Ocean.

(11)   The organo-volcanic spike and the equatorial crossing models represent two scenarios for the depositional environments of "anoxic" sediments in the Pacific Ocean, which do not have to employ an oxygen depletion of the oceanic intermediate and deep water on a global scale.

8.   Acknowledgments

This paper is based on samples and data collected by ourselves and by numerous colleagues who participated in the cruises of the D/V "Glomar Challenger" in the Pacific Ocean. The drill sites are all documented in the Initial Reports of the Deep-Sea Drilling Project. The authors acknowledge gratefully the effort of all of those who helped to make the Deep-Sea Drilling Project such a successful venture. This paper has benefited from the comments of H. C. Jenkyns (Oxford). This paper is a contribution to the IGCP Project "Mid-Cretaceous Events".

Arthur, M. A. & Natland, J. H. 1979. Carbonaceous sediments in the North and South Atlantic: The role of salinity in stable stratification of early Cretaceous basins. *Maurice Ewing Series* (American Geophysical Union) **3**, 375–401.

Boersma, A. 1981. Cretaceous and early Tertiary foraminifera from Leg 62 sites in the central Pacific. *Initial Reports of the Deep Sea Drilling Project* **62**, 377–396.

Byers, C. W. 1977. Biofacies patterns in euxinic basins: A general model. *Society of Economic Paleontologists and Mineralogists Special Publication* **25**, 5–17.

Cepek, P. 1981. Mesozoic calcareous nannoplankton stratigraphy of the central North Pacific (Mid-Pacific Mountains and Hess Rise), Leg 62. *Initial Reports of the Deep Sea Drilling Project* **62**, 397–418.

Curray, J., Moore, D. G. *et al.* 1979. Leg 64 seeks evidence on development of basins. *Geotimes* (July), 18–20.

Dean, W. E. 1981. Inorganic geochemistry and rocks from the Mid-Pacific Mountains and Hess Rise, Sites 463, 464, 465 and 466, Leg 62, Deep Sea Drilling Project. *Initial Reports of the Deep Sea Drilling Project* **62**, 685–710.

Dean, W. E., Claypool, G. E. & Thiede, J. 1981. Origin of organic carbon-rich mid-Cretaceous limestones, Mid-Pacific Mountains and southern Hess Rise. *Initial Reports of the Deep Sea Drilling Project* **62**, 877–890.

Dean, W. E. & Schreiber, B. C. 1978. Authigenic barite. Leg 41 Deep Sea Drilling Project. *Initial Reports of the Deep Sea Drilling Project* **41**, 915–931.

Degens, E. T. & Stoffers, P. 1976. Stratified waters as a key to the past. *Nature* **263**, 22–27.

de Graciansky, P.-C., Auffret, G. A., Dupeuble, P., Montadert, L. & Müller, C. 1979. Interpretation of depositional environments of the Aptian/Albian black shales on the north margin of the Bay of Biscay (DSDP Sites 400 and 402). *Initial Reports of the Deep Sea Drilling Project* **48**, 877–907.

Deroo, G., de Graciansky, P.-C., Habib, D. & Herbin, J.-P. 1978. L'origin de la matière organique dans les sédiments crétacés de site I.P.O.D. 398 (hautfond de Vigo): corrélations entre les données de la sédimentologie, de la géochemie organique et de la palynologie. *Société géologique de France Bulletin* **20**, 465–469.

Fischer, A. G. & Arthur, M. A. 1977. Secular variations in the pelagic realm. *Society of Economic Paleontologists and Mineralogists Special Publication* **25**, 19–50.

Gardner, J. V., Dean, W. E. & Vallier, T. L. 1980. Sedimentology and geochemistry of surface sediments, on the continental shelf, southern Bering Sea. *Marine Geology* **35**, 299–329.

Hallam, A. 1967. The depth significance of shales with bituminous laminae. *Marine Geology* **5**, 481–493.

Hamilton, E. L. 1956. Sunken islands of the Mid-Pacific Mountains. *Geological Society of America Memoir* **64**, 91pp.

Irving, E., North, F. K. & Coullard, R. 1974. Oil, climate and tectonics. *Canadian Journal of Earth Sciences* **11**, 1–17.

Jackson, E. D. & Schlanger, S. O. 1976. Regional Synthesis. Line Islands chain, and Manihiki Plateau, central Pacific Ocean. *Initial Reports of the Deep Sea Drilling Project* **33**, 915–927.

Jenkyns, H. C. 1980. Cretaceous anoxic events: from continents to oceans. *Journal of the Geological Society of London* **137**, 171–188.

Jenkyns, H. C. & Schlanger, S. O. 1981. Significance of plant remains in redeposited Aptian sediments, Hole 462A, Nauru Basin, to Cretaceous oceanic oxygenation models. *Initial Reports of the Deep Sea Drilling Project* **61**, 557–562.

Kemper, E. & Zimmerle, W. 1978. Die anoxischen Sedimente der präoberaptischen Unterkreide NW-Deutschlands und ihr paläogeographischer Rahmen. *Geologisches Jahrbuch* **A45**, 3–41..

Lancelot, Y. 1978. Rélations entre évolution sédimentaire et tectonique de la plaque pacifique depuis le Cretacée inférieur. *Société géologique de France Mémoirs N.S.* **134**, 39pp.

Larson, R. L., Moberly, R. *et al.* 1975a. Site 310: Hess Rise. *Initial Reports of the Deep Sea Drilling Project* **32**, 233–293.

Larson, R. L., Moberly, R. *et al.* 1975b. Site 305: Shatsky Rise. *Initial Reports of the Deep Sea Drilling Project* **32**, 75–158.

Larson, R. L., Moberly, R. *et al.* 1975c. Site 306: Shatsky Rise. *Initial Reports of the Deep Sea Drilling Project* **32**, 159–191.

Matthews, J. O. *et al.* 1974. Cretaceous drowning of reefs on Mid-Pacific and Japanese guyots. *Science*, **184**, 462–464.

Mélières, R., Deroo, G. & Herbin, J.-P. 1981. Organic-rich and hypersiliceous Aptian sediments from western Mid-Pacific Mountains, Deep Sea Drilling Project 62. *Initial Reports of the Deep Sea Drilling Project* **62**, 903–921.

Moore, T. C., Heath, G. R. & Kowsmann, R. O. 1973. Biogenic sediments of the Panama Basin. *Journal of Geology* **81**, 458–472.

Müller, P. J. & Suess, E. 1979. Productivity, sedimentation rate and sedimentary organic matter in the oceans—I. Organic carbon preservation. *Deep-Sea Research* **26**, 1347–1362.

Pitman, W. C., Larson, R. L. & Herron E. M. 1974. Magnetic lineations of the oceans. *Geological Society of America* Map.

Rea, D. & Harrsch, J. 1981. Mass accumulation rates of the non-authigenic inorganic crystalline (eolian) component of deep-sea sediments from the Hess Rise; DSDP Sites 464, 465 and 466. *Initial Reports of the Deep Sea Drilling Project* **62**, 661–668.

Rea, D. & Janecek, T. R. 1981. Mass accumulation rates of the non-authigenic crystalline (eolian) component of deep-sea sediments for the western Mid-Pacific Mountains, DSDP Site 463. *Initial Reports of the Deep Sea Drilling Project* **62**, 653–659.

Ryan, W. B. F. & Cita, M. B. 1977. Ignorance concerning episodes of ocean-wide stagnation. *Marine Geology* **23**, 197–215.

Sayre, W. O. 1981. Preliminary report on the paleomagnetism of Aptian and Albian limestones and trachytes from the Mid-Pacific Mountains and the Hess Rise. *Initial Reports of the Deep Sea Drilling Project* **62**, 983–994.

Schlanger, S. O., Jackson, E. D. *et al.* 1976. Site 317. *Initial Reports of the Deep Sea Drilling Project* **33**, 161–300.

Schlanger, S. O. & Jenkyns, H. C. 1976. Cretaceous oceanic anoxic events—causes and consequences. *Geologie en Mijnbouw* **55**, 179–184.

Schlanger, S. O., Jenkyns, H. C. & Premoli-Silva, I. 1981. Volcanism and vertical tectonics in the Pacific basin related to global Cretaceous transgression. *Earth and Planetary Science Letters 52*, 435–449.

Sclater, J. G., Anderson, R. N. & Bell, M. N. 1971. The elevation of ridges and evolution of the central eastern Pacific. *Journal of Geophysical Research* **76**, 7888–7915.

Seifert, K. E., Vallier, T. L., Windom, K. E. & Morgan, S. R. 1981. Geochemistry and petrology of DSDP Leg 62 igneous rocks. *Initial Reports of the Deep Sea Drilling Project* **62**, 945–953.

Simoneit, B. R. T. 1981. Organic geochemistry of Albian sediment from the Hess Rise, DSDP-IPOD Leg 62. *Initial Reports of the Deep Sea Drilling Project* **62**, 939–942.

Thiede, J. 1978. Pelagic sedimentation in immature oceanic basins. *Tectonics and geophysics of continental rifts* (Eds I. B. Ramberg & I. R. Neumann), Dordrecht, 237–248.

Thiede, J., Beorsma, A., Schmidt, R. R. & Vincent, E. 1981. Reworked fossils in Mesozoic and Cenozoic pelagic central Pacific Ocean sediments (Deep Sea Drilling Project sites 463, 464, 465 and 466, Leg 62). *Initial Reports of the Deep Sea Drilling Project* **62**, 495–512.

Thiede, J., Vallier, T. *et al.* 1981. Site 463: Mid-Pacific Mountains. *Initial Reports of the Deep Sea Drilling Project* **62**, 1073–1120.

Thiede, J. & van Andel, T. H. 1977. The paleoenvironment of anaerobic sediments in the late Mesozoic South Atlantic Ocean. *Earth and Planetary Science Letters* **33**, 301–309.

Timofeev, P. P. & Bogolyubova, L. I. 1981. Cretaceous sapropelic deposits of Sites 463, 465 and 466, DSDP Leg 62. *Initial Reports of the Deep Sea Drilling Project* **62**, 891–901.

Tissot, B., Durand, B., Espitalié, J. & Combaz, A. 1974. Influence of the nature and diagenesis of organic matter in formation of petroleum. *American Association of Petroleum Geologists Bulletin* **58**, 499–506.

Tissot, B., Eroo, G. & Herbin, J. P. 1979. Organic matter in Cretaceous sediments of the North Atlantic: Contribution to sedimentology and paleogeography. *Maurice Ewing Series* (American Geophysical Union) **3**, 362–374.

Vallier, T. *et al.* 1981. Hess Rise Synthesis. *Initial Reports of the Deep Sea Drilling Project* **62**, 1031–1072.

van Hinte, J. E. 1976. A Cretaceous time scale. *American Association of Petroleum Geologists Bulletin* **60**, 498–516.

Winterer, E. L., Ewing, J. I. *et al.* 1973. Site 171. *Initial Reports of the Deep Sea Drilling Project* **17**, 283–334.

Worsley, T. R. & Davies, T. A. 1979. Sea-level fluctuations and deep-sea sedimentation. *Science* **203**, 455–456.

5. Paleoenvironmental Significance of Clay Deposits in Atlantic Black Shales

H. Chamley

Sédimentologie et Géochimie, Université de Lille I, France

C. Robert

Géologie marine, Université d'Aix-Marseille II, France

Clay mineralogy investigations were performed on Cretaceous black shale sediments from various Deep Sea Drilling Project (DSDP) Sites, drilled on North and South Atlantic margins during DSDP Legs 11, 14, 36, 40, 41, 47B and 48. Well-crystallized smectite is the most common and abundant mineral. Associated clay minerals vary in type and abundance between DSDP Sites independently of lithology. Attapulgite occurs locally. These results, together with consideration of bulk mineralogy, organic and inorganic geochemistry, sedimentary structures and geographical distribution, point to a largely terrigenous origin for these Cretaceous black shales which differ in character from the Late Cenozoic Mediterranean sapropels. The detrital character of black clays allows their use in the reconstruction of past environmental conditions: conditions at the site of deposition favoured the preservation of allochthonous characteristics. The abundant Cretaceous smectite formed chiefly in badly drained coastal areas, a condition indicating tectonic stability and a low-relief continental morphology. The probable growth of smectite in gray–black soils points to the existence on the continents around the Atlantic Basin of a hot climate with strong contrasts in seasonal humidity. The various associated minerals generally reflect relatively local continental sources, bordering narrow ocean basins with weak north–south oceanic circulation. The Late Cretaceous diversification of terrigenous supply indicates the progressive development of longitudinal currents and marine water exchanges. Temporary increases of primary minerals, mixed-layers and kaolinite contents indicate marginal tectonics, related to global stages of ocean opening and widening.

1. Introduction

Researches on clay sedimentation in the Atlantic Basins led to a study of the facies of widespread Cretaceous black shales. At many sites of the Deep Sea Drilling Project, cores were recovered that permitted investigation of the character of clay deposition in these organic carbon-rich and associated sediments. Previous studies have been published in the *Initial Reports of the DSDP* or elsewhere (e.g. Lancelot *et al.*, 1972; Zimmerman, 1977; Chamley *et al.*, 1979a; Debrabant *et al.*, 1979; Mélières, 1979; Robert *et al.*, 1979).

Our purpose in this paper is to summarise the general patterns of clay sedimentation during the Cretaceous stages of black shale deposition in several Atlantic Basins. We apply mineralogical data to paleogeographical and paleoenvironmental interpretations in a parallel way to those who have used other methods (i.e. palynofacies: Habib, 1979; organic matter: Tissot *et al.*, 1979; conditions of circulation: Schlanger & Jenkyns, 1976; Ryan & Cita, 1977; Thiede & van Andel, 1977; Arthur, 1979; Thierstein, 1979). In addition we contrast the character of the

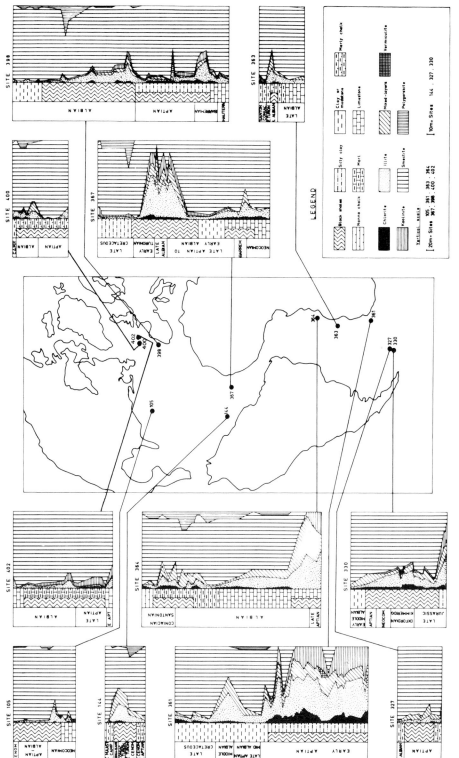

Figure 1. Site location in a Late Cretaceous geography, and clay successions identified in black shales and associated sediments.

Cretaceous black shales to the Plio–Pleistocene sapropels deposited in the Eastern Mediterranean.

Eleven DSDP sites were selected for study because of good core recovery of black shales and the under- and overlying sediments. Cores originate from Legs 11 (Site 105), 14 (Site 144), 36 (Sites 327 and 330), 40 (Sites 361, 363 and 364), 41 (Site 367), 47B (Site 398) and 48 (Sites 400 and 402), located in Eastern and Western Atlantic (Figure 1) (Hayes, Pimm *et al.*, 1972; Hollister, Ewing *et al.*, 1972; Barker, Dalziel *et al.*, 1976; Bolli, Ryan *et al.*, 1978; Lancelot, Seibold *et al.*, 1978; Sibuet, Ryan *et al.*, 1979a; Montadert, Roberts *et al.*, 1979). Detailed mineralogical studies are partly given in other publications (Chamley *et al.*, 1979a, 1980; Debrabant *et al.*, 1979; Roberts *et al.*, 1979).

The principle technique used was X-ray diffraction on the 2 μm decarbonated size fraction (see Chamley *et al.*, 1979a). Additional data was obtained by electronmicroscopy, inorganic (major traces and rare earth elements) and organic geochemistry, and lithologic studies.

2. Main features of clay sedimentation in Atlantic black shales— relationships with associated common sediments

Gray-black sediments, commonly called "black shales", occur in the Atlantic Basins in sediments of Late Mesozoic, chiefly Cretaceous age (Figures 1 and 2). In the Falkland area (South-Western Atlantic) they were deposited from Late Jurassic until Aptian time. In the North Atlantic they first appear in Hauterivian and persist until between Cenomanian and Santonian time. In the south-eastern Atlantic (from the Cape and Angola basins), black shales chiefly occur in Late Cretaceous strata only, because of a delay in the opening of these basins compared to the northern basins (e.g. Berggren & Hollister, 1977).

The lithological types existing in "black shales" facies are very diverse (Figure 1). They include chiefly gray to black mudstones and claystones which form heterogeneous successions of marly chalks, marls and graded sand–silt–clay sequences containing evidence of resedimentation. The conditions of deposition were very diverse, including suspended sediment regimes and mass-gravity flows; bottom depths were located below and above the calcite compensation depth (CCD), and strong variations in oxygen depletion of bottom waters occurred (see Arthur, 1979; de Graciansky & Chenet, 1979).

Smectite is the preponderant and most ubiquitous clay species encountered in the black shale facies, commonly forming 50–80% of the clay assemblage. Various other species occur also, generally rare but locally present in larger amounts. These include chlorite, illite, irregular mixed-layers of illite–smectite, chlorite–smectite and rarely illite–vermiculite types, vermiculite, kaolinite, palygorskite and sepiolite (Figure 1). Associated non-clay minerals chiefly include quartz, feldspars, pyrite, opal CT and clinoptilolite, diversely distributed in sediments. In a general way, the mineralogy of these clay assemblages differ quantitatively; changes can be correlated from site to site, and others are diachronous. The major trends are summarized in Figure 2, and concern two mineralogical groups. One group includes primary minerals (chlorite, illite), irregular mixed-layers and kaolinite; the other one comprises essentially smectite and fibrous clays (palygorskite, sepiolite).

There is no bed-to-bed relationship between the variations expressed by the clay mineralogy and by the lithology of the black shales and intercalated sediments. In the same way the clay assemblages do not depend on the abundance or nature of organic matter, and do not differ systematically in black shale facies and in under- and overlying sediments (Figure 3, right side). These data, combined with geochemical

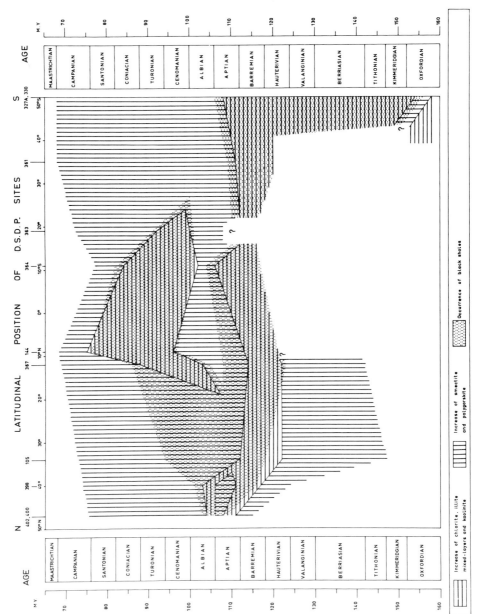

Figure 2.   Major clay mineralogical trends in Atlantic black shales facies and surrounding sediments.

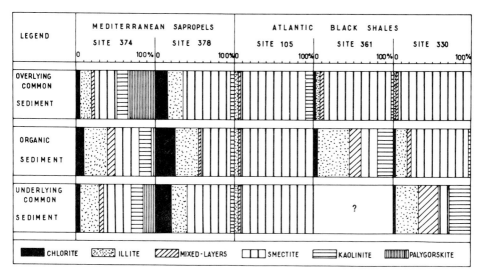

Figure 3. Behaviour of clay minerals in organic-rich and common sediments from Mediterranean (Quaternary) and Atlantic (Cretaceous).

results, point to the weakness or absence of *in situ* clay modifications related to the occurrence of black shales. The *in situ* noticeable modifications concern minerals less stable than clays, such as pyrite, clinoptilolite, opal-CT, as well as calcite, siderite, rodochrosite; these changes are not restricted to black shale facies (Debrabant *et al.*, 1979; Riech & von Rad, 1979). As a consequence, the Cretaceous environment leading to the deposition of black shales in the Atlantic realm seems to have caused little diagenesis of the clay minerals. This agrees with the evolution of organic matter, frequently marked by little *in situ* diagenesis and by low chemical reactivity (i.e. Deroo *et al.*, 1978; Tissot *et al.*, 1979).

The stability of clay assemblages in Cretaceous black shales strongly differs from the behaviour of these minerals in other organic carbon-rich sediments, such as the Plio-Pleistocene sapropels deposited in the Eastern Mediterranean. The latter consist of thin layers (from a few millimeters to one meter) with sharp bottom contacts. Mediterranean sapropels show strong diagenetic changes in clay mineralogy compared to the Cretaceous sediments: the most usual change in sapropels consists of a decrease of palygorskite and smectite, and a variable corrosion of phyllite particles. Figure 3 shows the average abundance of clay minerals in the sapropels, as well as in the sediment located immediately above and below the organic levels. The modifications, expressed by mineralogical, chemical and optical data, are synchronous and proportional to other sedimentary changes such as an increase in organic matter, amorphous silica (Diatom and Radiolarian debris), pyrite, trace elements and organo-mineral complexes and a decrease in carbonate content. The cause of the sediment modification is attributed to the reactivity of organic matter. Organic acids are responsible for the destruction of clays and carbonates, and for the concomitant development of other sedimentary components such as sulfur. Five stages of *in situ* clay diagenesis are recognized and interpreted in terms of increased destruction by organic acids in an euxinic environment at the sediment seawater interface. Only the first stages (partly destroyed clays) permit us to recognize the terrigenous sources and patterns of marine circulation during sapropelitic stages (Dominik and Stoffers, 1979). Most of the stages lead to the recognition of *in situ* modifications only which are characterized by transition minerals (chlorite, mixed-layers) and by successive mineralogical barriers against degradation (palygorskite, smectite, chlorite, illite, kaolinite). Results, discussions and references on this subject are given by Cita *et al.* (1977) and Sigl *et al.* (1978).

In summary, whereas the Mediterranean sapropelic environment determines important *in situ* synchronous modifications and appears to be chemically diversely reactive, the depositional environment of Atlantic black shales is much more passive and induces only minor chemical modifications. Clay materials, which are commonly very abundant in the black shale facies, do not come under noticeable bottom marine influence. As a consequence, clay assemblages of black shale facies may be used in reconstructing past conditions before their deposition.

3. Paleoenvironmental significance of clay assemblages in Atlantic black shales

3.1. *Sources of dominant clay materials*

The abundance and ubiquity of smectite not only characterizes the Cretaceous black shales and associated sediments, but occurs in the major part of the Late Mesozoic and Paleogene. Volcanogenic smectite is recognizable in altered basalts and in sediments a few centimetres above basaltic basement (Chamley & Bonnot-Courtois, 1981). Subaerial emission and alteration of basaltic materials favor the occurrence of abundant smectite, but such events are limited in time and space (i.e. Rockall Plateau during Eocene: Debrabant *et al.*, 1979); in oceanic sediments from DSDP Sites, numerous investigations point to the essentially terrigenous origin of smectite (i.e. Courtois & Chamley, 1978; Chamley, 1979*a*; Debrabant & Foulon, 1979). Smectite probably originated chiefly from continental soils, where the mineral formed in coastal and poorly-drained areas, as today in vertisoils (Pacquet, 1969). Smectite could also form partly in coastal sedimentary areas, marked by periodical emersions and conditions close to those of smectitic soils. The reworking of these surficial formations, unconsolidated, easily eroded and abundant in downstream areas, probably led to the massive supply of the minerals into the ocean (Chamley, 1979*a*). The natural blackish color of vertisoils kept on unaltered in reduced sediments such as black shales, marked by the preservative influence of organic matter. The preservation itself was aided by the relative narrowness and depth of the Atlantic Ocean, still favorable to deep water exchanges (i.e. Berggren & Hollister, 1977; Le Pichon *et al.*, 1978). In more oxidized sediments such as Early Cretaceous limestones (relatively shallow deposits) or Late Cretaceous varicolored clays (deep-sea deposition under active bottom circulation), the original color of smectitic soils disappeared, while iron oxides precipitated.

The main terrigenous origin of other clay minerals is obvious if we consider the lack of diagenesis with the depth of burial, the absence of a clear relationship between clay mineralogy and the lithologic facies encountered, and the diversity of clay assemblages which is similar to those deposited until recently in terrigenous environments. This assertion, evidenced by typical detrital minerals such as illite, chlorite and kaolinite, remains valid for most of Atlantic fibrous clays (palygorskite, sepiolite), inherited from coastal zones and diversely reworked on continental margins (see discussions in Timofeev *et al.*, 1978; Chamley, 1979*a*; Giblin, 1979).

3.2. *Climate*

The importance of detrital smectite supply to the ocean, when black shales and associated sediments are being deposited, implies subcontinuous conditions favoring the mineral formation in continental soils. Such conditions, revealed by geochemical studies on recent soils (i.e. Paquet, 1969), comprise especially specific climatic features, marked by high temperatures and humidity on emerged land-masses. These conditions are necessary for inducing strong hydrolyses of rocks in upstream, well-drained, continental areas, and the transfer of resulting ionic solutions towards the downstream areas. Warm and humid climate suggested by clay assemblages are also expressed by data on other sedimentary components (Furon, 1972; Reyre, 1979). Moreover the pedogenic development of smectite supposes irregular rainfall on coastal areas, leading to an alternating regime of wet and dry periods (Paquet, 1969; Chamley, 1979a). Such an alternating regime in continental humidity could be expressed in various time scales (from seasonal to pluriannual or pluricentenary cycles). Its existence agrees with the frequent abundance of organic matter directly reworked from land-masses (Habib, 1979; Tissot et al., 1979), with the common deposition of black shales in deep-sea fans developed during humid stages (de Graciansky & Chenet, 1979), and with the frequently general cyclic character of the sedimentation (Dean et al., 1979; Ferry et al., 1980).

The widespread distribution of pedogenically derived smectite in the Atlantic Ocean, independently of the lithologic nature of parent materials, suggests a large latitudinal extension of hot and humidity-contrasted climate on American and Euro-African land-masses. The latitudinal climatic zonation during black shale deposition was still marked, probably due to the narrowness of the young ocean, the probable absence of glaciation and the weakness of longitudinal water exchanges. Climatic zonation in continental soils and clays took place much later, during the latest Cretaceous widening of Atlantic, followed by the stepwise Cenozoic cooling (Chamley, 1979a; Chamley & Robert, 1979).

3.3. Morphology

The morphology of adjacent emerging areas influenced the composition of detrital assemblages supplied to ocean during black shale deposition. The smectite pedogenic development supposes a poor circulation of ground water in surficial formations, allowing ionic trapping and mineral growth (i.e. Paquet, 1969). Such conditions develop only in flat areas with poor water drainage. As a consequence, the coastal morphology during black shale deposition was probably marked by low-relief and an evoluted topography, upon which developed blackish soils and periodically abundant hydromorphic vegetation. This situation, often extending to preceding and following periods, supposes a general preponderance of quiescent stages over tectonic stages, the latter developing temporarily only (de Graciansky et al., 1978; Robert et al., 1979; Chamley et al., 1980).

Peculiar morphologic conditions existed during palygorskite-rich stages, locally and episodically occurring when black shales were deposited, especially at Albian and Coniacian–Santonian times (Figure 1). The formation of sedimentary palygor-skite implies hot and humidity-contrasted climate, as is the case for smectite formation. But additional conditions consist in the existence of closed to semi-closed basins, often associated with a regime of marine transgression, and favoring both ionic supply and strong water concentrations (i.e. Millot, 1964; Trauth, 1977; Weaver & Beck, 1977; Chamley, 1979a). Such basins were present and geochemically active on Atlantic margins at several periods from Middle–Late Mesozoic to Paleogene, and are not restricted to the periods of black shale development. The formation of fibrous clays in marginal basins depended on a complex environmental equilibrium including climatic, eustatic and morphologic influences. The mor-phology of coastal areas, linked to the eustacy regime (Cooper, 1977), seems to have been the outstanding factor inducing the development of fibrous clay stages, climate appearing to have been suitable throughout the period concerned (see above). This points to the lack of stratigraphic significance of fibrous clay development, the occurrence of favorable marginal topography often having a regional character and not a general one; only a few periods, such as Early Eocene time, seem to have a more general extension (i.e. Millot, 1964; Chamley, 1979a). The reworking of fibrous clays, from marginal basins to deep open marine environment where the DSDP holes were drilled, benefited by the chronic instability of Atlantic margins reacting to sub-permanent spreading (faulting, uplift or subsidence).

3.4. Oceanic circulation

The development of Mesozoic black shales was determined by the existence of a young Atlantic ocean, already relatively deep but still subject to restricted oceanic circulation, especially in deep and intermediate water masses (e.g. Schlanger & Jenkyns, 1976; Berggren & Hollister, 1977; Thiede & van Andel, 1977; Arthur, 1979). This peculiar situation led to specific dispersion conditions of terrigenous materials supplied to the ocean, namely detrital clay minerals. Local and regional supplies predominated over distant and general supplies, which is clearly illustrated by the mineralogical developments characteristic of each drill site (chlorite, illite, mixed-layers, kaolinite or palygorskite: see Figure 1). This relationship between the occurrence of restricted basins and the predominance of local terrigenous supplies decreased later on, when the widening of the Atlantic ocean favored the development of north–south exchanges (Chamley, 1979a; Robert, 1980).

3.5. Tectonic activity

The dominant sedimentation of clay assemblages with very abundant smectite during black shale deposition was interrupted by relatively short but strong supplies of other clay minerals (Figure 1). The supply of primary minerals and mixed-layers increased. These abundant and sudden supplies are interpreted as markers of strong erosional periods on adjacent continental margins, favoring the removal of rocks and slowing the formation of soils. Other minerals increased only locally and temporarily, depending on the morphology and climate of adjacent land-masses (i.e. Aptian kaolinite in Cape and Angola basins, Sites 361 and 364; Albian palygorskite off the Iberian and Armorican peninsulae, Site 402; Albo-Turonian kaolinite in the Cape Verde basin, Site 367). All these short-term mineralogical events, marked by a strong decrease in smectite abundance and crystallinity, and by an increase and a diversification of various other clay species, principally illite and mixed-layers, are interpreted as evidence of tectonic activity on adjacent land-masses (Chamley, 1979b; Chamley et al., 1979b; Robert et al., 1979). Ordinarily these mineralogical interruptions of the smectitic background reflect structural reactions of Atlantic margins to the successive stages of ocean opening and widening. Three examples of such reactions appear during the periods of black shale deposition (Figures 1 and 2).
3.5.1.   In the South-East Atlantic (Cape basin, Site 361), massive output of primary minerals and kaolinite occurred at Early Aptian time, obliterating the supply of smectite as a result of the tectonic activation of the Africa margin (Robert et al., 1979; Maillot & Robert, 1980). The same event exists with a slight time shift from South towards the North: its manifestations appear during Late Aptian–Early Albian in Angola basin, and in Albian to Cenomanian gray deposits from Gabon and Nigeria basins (Chamley et al., 1979b; Maillot & Robert, 1980, and unpubl. res.). The delay is attributed to morphological differences between the southern and northern parts of the South Atlantic, the former being largely open a long time before the latter. Note that in the North Atlantic, similar tectonic events occurred a long time

before Cretaceous time and black shale deposition, because of an earlier chronology of ocean opening than in South Atlantic (in Chamley, 1979a). As far as the most southern part of the Atlantic area is concerned (Falkland Plateau, Site 330, Figure 1), the precocious development of primary minerals supply is related to the existence of an older history, e.g. the Jurassic opening of the Southern Ocean (Barker, Dalziel *et al.*, 1977).

3.5.2.   In the North Atlantic, the black shale deposition is contemporary to less intense but is marked by more numerous obliterations of pedogenic smectite formation (i.e. during Barremian to Albian in the Biscay Bay and Cape Hatteras basins at Sites 398, 400, 402, 105). The temporary floods of illite and associated minerals are considered as reflections of successive major phases of spreading in the Atlantic, marked on its continental margins by a morphological rejuvenation (Chamley, 1979a; Roberts & Montadert, 1979; Sibuet & Ryan, 1979b). In the South Atlantic similar minor tectonic events expressed by clay associations occurred after the periods of black shale deposition.

3.5.3.   In the Central Atlantic, a major mineralogical event is recorded during the Late Cretaceous by the appearance of an increase in chlorite, illite and mixed-layer abundance from Late Albian to Turonian time in the Cape Verde basin (Site 367), on the Demerara Rise (Site 144), in the Angola basin (Site 364) and on the northern flank of the Walvis Ridge (Site 363). This mineralogic change is attributed to a tectonic event chiefly located close to the Vema-Romanche fracture zones and in the Benue Trough. The tectonic activity, marked by a rejuvenation of continental margins and by an increase in the intensity of oceanic volcanism, probably results from a major spreading stage leading to the Late Cretaceous separation of Africa and South America (Berggren & Hollister, 1977; Chamley, 1979b; Thiede, 1979; Chamley *et al.*, 1980).

In a more general way, rough relationships frequently exist between the development of the black shale facies and the increase of illite and associated minerals (Figure 2). Tucholke and Vogt's geological observations (1979) point to the existence of a synchronism between the development of black shales and tectonic activity (margin movements, inset of submarine barriers, etc.). During Late Jurassic and Early Cretaceous, the diminution of global oceanic circulation induced a stagnation of deep waters (Arthur, 1979). By this time, young oceanic basins formed along the periphery of the Tethyan and Southern oceans, in relation with marginal tectonic activity. The conjunction of reduced circulation and narrow oceanic basins may be responsible for the origin of the black shales. As a consequence, clay minerals appear to express the tectonic activity related to the morphologic evolution of the basins, while organic deposits reflect the evolution of the euxinic marine environment induced by the restricted circulation.

Subsequent events consist of continental relaxation and topographical planation, propitious to the development of smectite and often synchronous with more oxygenated oceanic conditions. A second stage of both black shale deposition and primary mineral supply occurs in Late Cretaceous time in the Angola Basin and the Walvis Ridge, resulting from similar structural changes (ocean deepening and continental reaction linked to the opening of the Vema–Romanche fracture zone and Benue Trough), and ending when strong meridian currents developed (McCoy & Zimmerman, 1977; Le Pichon *et al.*, 1978).

4. Conclusion

Data and interpretations from selected DSDP sites led to a review of the main patterns of clay sedimentation during the deposition of Mesozoic black shales and associated sediments in the Atlantic Ocean. The following main observations can be made.

(1) The black shale marine environment induces only minor chemical modifications of sedimentary components, in contrast to the marked diagenesis seen in the Mediterranean Plio-Pleistocene sapropels.

(2) As a consequence, the study of detrital materials permits the deciphering of the paleoenvironmental conditions prevailing on land-masses adjacent to the Atlantic basins. Clay minerals are largely derived from both continental rocks and soils.

(3) The background of clay sedimentation, at the time of black shale deposition, consists of an abundant supply of well-crystallized smectite, chiefly born in coastal soils and pointing to the conjunction of an average hot climate with strong alternations in humidity with flat coastal topography and tectonic quiescence.

(4) Successive overriding of smectitic sedimentation by various other clay assemblages reflect the occurrence of paleogeographic events. The most important are the following:short-term existence of peri-marine semi-closed to closed basins, often linked to a transgressive regime and to strong contrasts in humidity conditions, which led to the development of fibrous clay assemblages periodically reworked towards the open ocean by marginal subsidence; tectonic rejuvenation stages, evidenced by strong and temporary supplies of primary and associated minerals, and determined by successive phases of ocean formation and widening as follows: Early Cretaceous phases of South Atlantic Opening; Barremian to Albian spreading phases of North Atlantic; Albian to Turonian separation of Africa and America, expressed in the central ocean.

(5) The appearance of black shales in the stratigraphic record in the Atlantic roughly corresponds to increases in the supply of illite and associated minerals, especially in the South Atlantic whose history is relatively simple. This correspondence probably results from tectonic events, which determine both oceanic deepening and margin rejuvenation. Other relationships between black shale deposition and clay mineralogy concern the predominance of local and regional terrigenous supplies relative to those from distant sources. Black shale development is also linked to the relative narrowness and shallowness of the early Atlantic Basins.

Acknowledgments

The National Science Foundation (U.S.A.) has allowed us since 1976 to study many DSDP materials. The financial support was provided by CNEXO (grants No. 76/5320, 77/5155, 78/5708, 79/5927), CNRS (ATP IPOD 1977) and DGRST (No. 78-7-2 941). Technical assistance was given by M. Acquaviva, M. Bocquet, J. Carpentier, F. Dujardin and C. H. Froget.

References

Arthur, M. A. 1979. North Atlantic Cretaceous black shales: the record at Site 398 and a brief comparison with other occurrences. *Initial Reports of the Deep Sea Drilling Project* **47** (2), 719–751.
Barker, P. F., Dalziel, I. W. D. *et al.* 1976. *Initial Reports of the Deep Sea Drilling Project* **36** (2), 1–1080.

Berggren, W. A. & Hollister, C. D. 1977. Plate tectonics and paleocirculation. Commotion in the Ocean. *Tectonophysics* **38**, 11–48.

Bolli, H. M., Ryan, W. B. F. *et al.* 1978. *Initial Reports of the Deep Sea Drilling Project* **40**, 1–1067.

Chamley, H. 1979a. North Atlantic clay sedimentation and paleoenvironment since the Late Jurassic. *Deep Drilling Results in the Atlantic Ocean: Continental Margins and Paleoenvironment* (Eds M. Talwani, W. Hay & W. B. F. Ryan), Maurice Ewing Series 3, *American Geophysical Union Publication* 342–361.

Chamley, H. 1979b. Les successions argileuses de L'Atlantique Nord, écho des changements mésozoïques et cénozoïques de l'environnement. Exemple du bassin du Cap Vert. *Comptes Rendus de Academie Sciences Paris* **289**-D, 769–772.

Chamley, H., Bonnot-Courtois, C. 1981. Argiles authigènes et terrigènes de l'Atlantique et du Pacifique NW (legs 11 et 58 DSDP): apport des terres rares. *Oceanologica Acta* **4** (2), 229–339.

Chamley, H., Debrabant, P., Foulon, J., Giroud d'Argoud, G., Latouche, C., Maillet, N., Maillot, H. & Sommer, F. 1979a. Mineralogy and Geochemistry of Cretaceous and Cenozoic Atlantic sediments off the Iberian Peninsula (site 398, DSDP Leg 47B) *Initial Reports of the Deep Sea Drilling Project* **47** (2), 429–449.

Chamley, H., Debrabant, P., Foulon, J. & Leroy, P. 1980. Contribution de la minéralogie et de la géochimie à l'histoire méso-cénozoïque des marges nord-atlantiques (sites 105 et 367 DSDP). *Bull. Soc. géol. France* **22** (7), 745–755.

Chamley, H., Enu, E., Moullade, M. & Robert, C. 1979b. La sédimentation argileuse du bassin de la Bénoué au Nigéria, reflet de la tectonique du Crétacé supérieur. *Comptes Rendus de l'Academie Sciences Paris* **288**-D (1), 143–146.

Chamley, H. & Robert, C. 1979. Late Cretaceous to early Paleogene environmental evolution expressed by the Atlantic clay sedimentation. *Cretaceous–Tertiary Boundary Events Symposium* (Eds W. K. Christensen and T. Birkelund), Vol. 2. Proceedings. University of Copenhagen. pp. 71–77.

Cita, M. B., Vergnaud-Grazzini, C., Robert, C., Chamley, H., Ciaranfi, N. & D'Onofrio, S. 1977. Paleoclimatic record of a long deep sea core from the Eastern Mediterranean. *Quaternary Research* **8**, 205–235.

Cooper, M. R. 1977. Eustacy during Cretaceous; its implications and importance. *Palaeogeography, Palaeoclimatology, Palaeoecology* **22**, 1–60.

Courtois, C. & Chamley, H. 1978. Terres rares et minéraux argileux dans le Crétacé et le Cénozoïque de la marge atlantique orientale. *Comptes Rendus l'Academie Sciences Paris* **286**-D, 671–674.

Dean, W. E., Gardner, J. V., Jansa, L. F., Cepek, P. & Seibold, E. 1979. Cyclic sedimentation along the continental margin of Northwest Africa. *Initial Reports of the Deep Sea Drilling Project* **41**, 965–989.

Debrabant, P., Chamley, H., Foulon, J. & Maillot, H. 1979. Mineralogy and geochemistry of upper Cretaceous and Cenozoic sediments from North Biscay Bay and Rockall Plateau (Eastern North Atlantic), DSDP Leg 48. *Initial Reports of Deep Sea Drilling Project* **48**, 703–725.

Debrabant, P. & Foulon, J. 1979. Expression géochimique des variations du paléoenvironment depuis le Jurassique supérieur sur les marges nord-atlantiques. *Oceanologica Acta* **2** (4), 469–476.

de Graciansky, P. Ch. & Chenet, P. Y. 1979. Sedimentological study of cores 138 to 56 (upper Hauterivain to middle Cenomanian): an attempt at reconstruction of paleoenvironments. *Initial Reports of the Deep Sea Drilling Project* **47** (2), 403–418.

de Graciansky, P. Ch., Müller, C., Rehault, J. P. & Sigal, J. 1978. Reconstitution de l'évolution des milieux de sédimentation sur la marge continentale ibérique au Crétacé: le flanc sud du haut-fond de Vigo et le forage D.S.D.P. I.P.O.D. 398. Problèmes concernant la surface de compensation des carbonates. *Bull. Soc. géol. France* **20** (7), 389–400.

Deroo, G., de Graciansky, P. Ch., Habib, D. & Herbin, J. P. 1978. L'origine de la matière organique dans les sédiments crétacés du site I.P.O.D. 398 (haut fond de Vigo): corrélations entre les données de la sédimentologie, de la géochimie organique et de la palynologie. *Bull. Soc. géol. France* **20** (4), 465–471.

Dominik, J. & Stoffers, P. 1979. The influence of late Quaternary stagnations on clay sedimentation in the Eastern Mediterranean Sea. *Geol. Rundschau* **68**, 302–317.

Ferry, S., Cotillon, P., Gaillard, C., Sautée, E., Latreille, G. & Rio, M. 1980. Alternances marno-calcaires: bruit de fond universel de la sédimentation pélagique au-dessus de la profondeur de compensation des carbonates. *8ème R.A.S.T.*, Marseille, 148pp.

Furon, R. 1972. *Eléments de paléoclimatologie.* Vuibert ed., Paris, 1–216.

Giblin, P. 1979. Minéralogie et géochimie de la limite Crétacé-Tertiaire dans quelques forages océaniques profonds. *Thèse 3ème cycle*, Strasbourg, 1–99.

Habib, D. 1979. Sedimentary origin of North Atlantic Cretaceous palynofacies. *Deep Drilling Results in the Atlantic Ocean: Continental Margins and Paleoenvironment* (Eds M. Talwani, W. Hay & W. B. F. Ryan), Maurice Ewing Series 3, *American Geophysical Union Publication*, 420–437.

Hayes, D. E., Pimm, A. C. *et al.* 1972. *Initial Reports of the Deep Sea Drilling Project* **14**, 1–975.

Hollister, C. D., Ewing, J. T. *et al.* 1972. *Initial Reports of the Deep Sea Drilling Project* **11**, 1–1077.

Lancelot, Y., Hathaway, J. C. & Hollister, C. D. 1972. Lithology of sediments from the Western North Atlantic, Leg 11, Deep Sea Drilling Project. *Initial Reports of the Deep Sea Drilling Project* **11**, 901–949.

Lancelot, Y., Seibold, E. *et al.* 1976. *Initial Reports of the Deep Sea Drilling Project* **41**, 1–1259.

Le Pichon, X., Melguen, M. & Sibuet, J.-C. 1978. A schematic model of the evolution of the South Atlantic. *Advances in oceanography* (Eds H. Charnock & S. G. Deacon), Plenum, 1–48.

Maillot, H. & Robert, C. 1980. Minéralogie et géochimie des sédiments Crétacés et cénozoïques dans l'Océan Atlantique Sud (marge africaine, dorsale médio-atlantique). *Bull. Soc. géol. France* **22** (7), 777–788.

McCoy, F. W. & Zimmerman, H. B. 1977. A history of sediment lithofacies in the South Atlantic Ocean. *Initial Reports of the Deep Sea Drilling Project* **39**, 1047–1080.

Mélières, F. 1979. Mineralogy and geochemistry of selected Albian sediments from the bay of Biscay, Deep Sea Drilling Project Leg 48. *Initial Reports of the Deep Sea Drilling Project* **48**, 855–875.

Millot, G. 1964. *Géologie des argiles*. Masson ed., Paris, 1–499.

Montadert, L., Roberts, D. G. *et al.* 1979. *Initial Reports of the Deep Sea Drilling Project* **47** (2), 1–787.

Paquet, H. 1969. Evolution géochimique des minéraux argileux dans les altérations et les sols des climats méditerranéens et tropicaux à saisons contrastées. *Sci. Géol.*, Strasbourg, Mém. 30, 1–210.

Reyre, Y. 1979. L'évolution climatique au Jurassique et au Crétacé vue à travers la palynologie. *IVème Symposium Ass. Palynologues de Langue Française Palynologie et climats*. Entretiens du Muséum, Résumés des communications 32.

Riech, V. & von Rad, U. 1979. Silica diagenesis in the Atlantic Ocean: diagenetical potential and transformation. *Deep Drilling Results in the Atlantic Ocean: Continental Margins and Paleoenvironment* (Eds M. Talwani, W. Hay & W. B. F. Ryan). Maurice Ewing Series 3, *American Geophysical Union Publication*, 315–340.

Robert, C. 1980. Climats et courants cénozoïques dans l'Atlantique Sud d'après l'étude des minéraux argileux (Legs 3, 39 et 40 DSDP). *Oceanologica Acta* (in press).

Robert, C., Herbin, J.-P., Deroo, G., Giroud d'Argoud, G. & Chamley, H. 1979. L'Atlantique Sud au Crétacé d'après l'étude des minéraux argileux et de la matière organique (Legs 39 et 40 DSDP). *Oceanologica Acta* **2**, 209–218.

Roberts, D. G. & Montadert, L. 1979. Evolution of passive rifted margins. Perspective and retrospective of DSDP Leg 48. *Initial Reports of the Deep Sea Drilling Project* **48**, 1143–1153.

Ryan, W. B. F. & Cita, M. B. 1977. Ignorance concerning episodes of ocean-wide stagnation. *Marine Geology* **23**, 197–215.

Schlanger, S. O. & Jenkyns, H. C. 1976. Cretaceous oceanic anoxic events: causes and consequences. *Geologie en Mijnbouw* **55**, 179–184.

Sibuet, J.-C., Ryan, W. B. F. *et al.* 1979*a*. *Initial Reports of the Deep Sea Drilling Project* **47** (2), 1–1183.

Sibuet, J.-C. & Ryan, W. B. F. 1979*b*. Site 398: Evolution of the West Iberian passive continental margin in the framework of the early evolution of the North Atlantic Ocean. *Initial Reports of the Deep Sea Drilling Project* **47** (2), 761–787.

Sigl, W., Chamley, H., Fabricius, F., Giroud d'Argoud, G. & Müller, J. 1978. Sedimentology and environmental conditions of sapropels. *Initial Reports of the Deep Sea Drilling Project* **42**, 445–464.

Thiede, J. 1979. Paleogeography and paleobathymetry of the Mesozoic and Cenozoic North Atlantic Ocean. *Geojournal* 3, 263–272.

Thiede, J. & van Andel, T. H. 1977. The paleoenvironment of anaerobic sediments in the late Mesozoic South Atlantic Ocean. *Earth Planet. Sci. Letters* **33**, 301–309.

Thierstein, H. R. 1979. Paleoceanographic implications of organic carbon and carbonate distribution in Mesozoic deepsea sediments. *Deep Drilling Results in the Atlantic Ocean: Continental Margins and Paleoenvironment* (Eds M. Talwani, W. Hay & W. B. F. Ryan), Maurice Ewing Series 3, *American Geophysical Union Publication*, 249–274.

Timofeev, P. P., Eremeev, V. V. & Rateev, M. A. 1978. Palygorskite, sepiolite and other clay minerals in Leg 41 oceanic sediments: Mineralogy, facies and genesis. *Initial Reports of the Deep Sea Drilling Project* **41**, 1087–1101.

Tissot, B., Deroo, G. & Herbin, J. P. 1979. Organic matter in Cretaceous sediments of the North Atlantic: Contribution to sedimentology and paleogeography. *Deep Drilling Results in the Atlantic Ocean: Continental Margins and Paleoenvironment* (Eds M. Talwani, W. Hay & W. B. F. Ryan), Maurice Ewing Series 3, *American Geophysical Union Publication*, 362–374.

Trauth, N. 1977. Argiles évaporitiques dans la sédimentation carbonatée tertiaire. Bassins de Paris, de Mormoiron et de Salinelles (France); Jbel Ghassoul (Maroc). *Sci. géol. Strasbourg* **49**, 1–203.

Tucholke, B. E., Vogt, P. R. 1979. Western North Atlantic: sedimentary evolution and aspects of tectonic evolution. *Initial Reports of the Deep Sea Drilling Project* **43**, 791–825.

Weaver, C. E. & Beck, K. C. 1977. Miocene of the S-E United States: a model for chemical sedimentation in a peri-marine environment. *Sedimentary Geology* **17** (1/2 spec. issue), 1–234.

Zimmerman, A. B. 1977. Clay mineral stratigraphy and distribution in the South Atlantic Ocean. *Initial Reports of the Deep Sea Drilling Project* **39**, 395–405.

6. Sedimentary Supply Origin of Cretaceous Black Shales

D. Habib

Department of Earth and Environmental Sciences, Queens College, New York

A hypothesis based on the sedimentary supply of organic matter is proposed for the origin of Cretaceous black shales in the North Atlantic. Early Cretaceous deltaic systems episodically contributed large amounts of terrigenous organic detritus, which was sorted and redistributed in a direction away from source vegetation. Sea-surface fertility and organic productivity in both the marginal seas and open ocean contributed organic matter to marine carbon-rich black shales during the maximum transgression of early Late Cretaceous age. The supply of organic matter and its rate of burial were the dominant factors controlling the kind and distribution of organic carbon in palynologically-defined organic facies. High rates of sedimentation carried abundant diversified and well-preserved organic matter quickly through the diagenetic zone of aerobic oxidation. Lower rates exposed organic matter to aerobic oxidation. Models which require anoxic bottom waters as a prerequisite for the preservation of organic matter are not necessary to explain the occurrence of either organic-rich or organic-poor black shales; the bottom waters could have been oxidizing.

1. Introduction

The North Atlantic Lower Cretaceous is characterized by blackish-colored, generally organic-rich, sediments containing organic matter primarily of land-plant origin. An organic-walled microflora (dinoflagellates, small acritarchs) occurs also, and is an important constituent in several intervals of the sedimentary column, but is not a predominant component until the Upper Albian or Upper Cretaceous. The origin of these carbonaceous sediments has been a major topic of debate, at least since the study of the bathymetrically deep submarine outcrop of black shales of Albian age located east of Cat Island in the Bahamas (Windisch *et al.*, 1968; Habib, 1968) and subsequently from the studies of numerous sections recovered through the drilling programs undertaken by the Deep Sea Drilling Project. A number of hypotheses have been offered to reconstruct the environmental conditions under which this organic matter accumulated. Most require anoxic bottom waters as a necessary prerequisite for the preservation of organic matter, caused either by stagnant conditions developed in a stratified Atlantic Ocean (Black Sea model discussed in Degens & Ross (1974) or injection of saline waters from the opening South Atlantic proposed by Thierstein & Berger (1978)) or by thermohaline-related vertical expansion of an oxygen-minimum zone in a deep water column which intermittently contacted bottom surface sediments (Fischer & Arthur, 1976). These hypotheses emphasize preservation of organic matter as the method of enriching the organic carbon content of bottom sediments, regardless of its rate of supply from the land surface bordering the North Atlantic or the productivity of marine organisms (Arthur & Natland, 1979).

The purpose of this study is to present palynological evidence in support of the hypothesis that the carbonaceous character of black claystones and black clay-laminated mudstones and limestones (collectively called black shales), both organic carbon-rich and organic carbon-poor, are the direct results of the kinds and rates of organic matter supplied to the North Atlantic sea floor during the Early Cretaceous and earlier Late Cretaceous. The rate of burial of organic matter is considered to have been sufficiently high to have caused anoxic conditions in the early diagenetic environment regardless of the conditions prevalent in the overlying bottom waters, oxygen-rich or oxygen-depleted (Curtis, 1980). The sedimentation of organic matter is considered to have been related to processes operating at or very near the continental margins, with respect to the influx of deltaic sediments episodically during much of the Early Cretaceous and to the migration of shorelines away from the central North Atlantic during the geographically extensive marine transgressions of Albian to early Late Cretaceous times. The sedimentation of organic matter of both terrestrial and marine origin is considered to have controlled the amount and kind of organic carbon as well as the color of these sediments.

During the Early Cretaceous, the continental shelves bordering the North Atlantic were aggrading as the direct consequence of massive deltaic systems (Ryan & Cita, 1977). The preserved deltaic facies is regressive and widespread, ranging in outcrop and in subsurface sections from the eastern margin of the United States (Minard et al., 1974) and south-eastern Canada (Jansa & Wade, 1975) across to the Weald delta of England and western Europe (Allen, 1969), and south to the margin of western North Africa (Einsele & von Rad, 1979). Numerous wells drilled by petroleum companies attest to the extent of the Lower Cretaceous deltaic facies which virtually ringed the Cretaceous North Atlantic. Study of the C.O.S.T. B-2 well, drilled in the Baltimore Canyon at the outer margin of the U.S. Atlantic continental shelf, shows a thick section of prograding deltaic facies of Barremian–Early Albian age, characterized by abundant terrestrial organic matter (personal observation, and Report No. EPGS-PAL-6-78GLW distributed by the Bedford Institute of Oceanography).

This facies is palynologically rich. The abundance and diversity of plant fragments of terrestrial origin, including pollen grains, fern spores, woody tissue (tracheids), and epidermal cuticles reflects the richness of the vascular plant life which occupied the backswamp areas of the deltas. The outpouring of these extensive systems, very close to the outer limits of the present continental margins, into the Early Cretaceous ocean contributed very large amounts of terrigenous organic sediments to the bottom environments.

During the middle Cretaceous (Albian–Cenomanian) the North Atlantic transgressed the surrounding continents, submerging previously emergent shelves and flooding deltaic sediments as the shorelines moved landward. This phase culminated in the maximum transgression of Cenomanian age, which affected the continents bordering the Cretaceous North Atlantic (Gignoux, 1955).

2. Sedimentary model

A sedimentary model is proposed to explain the stratigraphic and geographic distribution of organic matter in ten sections drilled by the Deep Sea Drilling Project (Figure 1). The model is based on the premise that deltaic systems supplied the bulk of organic matter contained in these sections during episodes of increased progradation, and that sedimentary processes operating in the ocean sorted and otherwise modified the material away from the major sources of vegetation. A study of three sections recovered at Sites 105, 391C and 398D (Figure 1) has been published previously (Habib, 1979a). The results of this study and that of seven additional sections, recovered at Sites 99A, 101A, 135, 136, 386, 387 and 397A, are integrated in order to determine if geographic trends away from the margins are the same as those originally proposed.

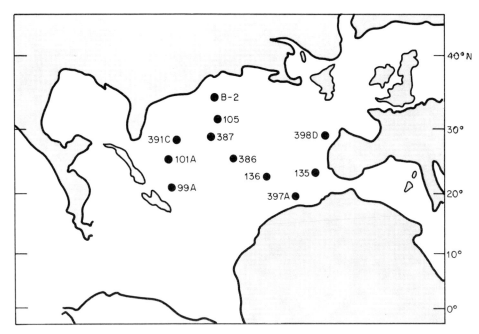

Figure 1.   Generalized reconstruction of Albian North Atlantic, showing distribution of the C.O.S.T. B-
2 well and ten D.S.D.P. sites studied.

Figure 2 is a schematic representation of the distribution of the various
sedimentary organic particles. Various categories of particles are distinguished and
related to each other in defined organic facies (palynofacies of Habib, 1979b). The
horizontal lines represent the range of each category where it is most abundant.
Thus, fern spores larger than 50 μm in diameter, and those attributed to the family
Schizaeaceae, are always most abundant where the rate of sedimentation of organic
matter is highest, such as within prodeltaic turbidites (Site 398D) and prodeltaic
laminated mudstones (Site 397A). Larger fern spores are also found where there is
diminished terrigenous sedimentation, but they are extremely rare. The organic
facies are correlated with sedimentary parameters such as amount of terrigenous
sediment, estimated rates of sedimentation, sedimentary structures such as turbidites, and the locus of the facies relative to Cretaceous shorelines.
    As illustrated in Figure 2, the exinitic facies consists of numerous pollen grains,
large fern spores, tracheids, and cuticles (in the order of 60 000–100 000 specimens
per gram), and the almost total absence of dinoflagellates. The preservation of the
organic matter ranges from very good to excellent, with some spores appearing as
fresh as modern material (Habib, 1979a, pl. 1). Carbonized tracheids occur, together
with minor amorphous debris, but these are larger-sized and not as numerous. The
exinitic facies occurs in sections (Sites 397A, 398D) closest to the continental
margins in intervals of very high rates of sedimentation (50–100 m per m.y.), and is
considered to be the direct result of prodeltaic processes. Its composition is very
close to that occurring in the fluvio-deltaic Potomac Group of Maryland, in the
C.O.S.T. B-2 well, as well as in the Wealden facies of England (Batten, 1973).
    The tracheal facies also consists of abundant terrigenous materials. It is essentially
contemporaneous with the exinitic facies but was formed in sections (Sites 101A,
105, 135, 391C) farther from a source of vegetation (Figure 3). Pollen grains remain
numerically abundant (Figure 2), in the order of 40 000–60 000 per gram, but they are
now dominated by those morphotypes (bisaccates and small spheroidal grains in

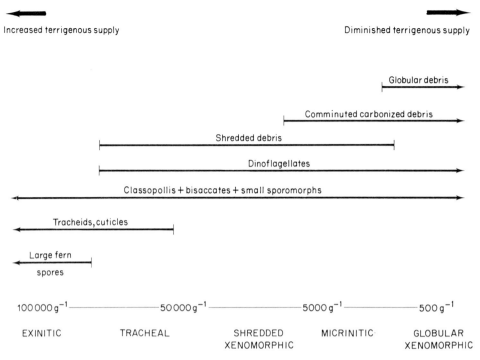

Figure 2.   Sedimentary model depicting admixtures of more abundant structured particulate organic matter in palynologically-defined organic facies. Numbers represent specimens of pollen grains, spores, dinoflagellates, and tracheids per gram of sediment. Abundant amorphous debris is estimated, as it ranges to millions of particles per gram.

Classopollis) adapted for longer-distance transportation (Habib, 1979*b*). Large fern spores are infrequent. Tracheids remain abundant, but there now is an admixture of dinoflagellate cysts. The rate of sedimentation in the tracheal facies is considerably less, but in the intervals where it is expressed, it is considered to be higher than the 10–30 m per m.y. average for the entire Lower Cretaceous published for the investigated sections. The lithology containing the tracheal facies is primarily carbonaceous clay-laminated nannofossil limestone. However, these intervals contain numerous graded layers indicative of terrigenous turbidite deposition (Benson, Sheridan *et al.*, 1978; Ryan & Cita, 1977, p. 201), although on a smaller scale.

In contrast to the above, the remaining organic facies are defined by an abundance of various kinds of amorphous debris (shredded, carbonized, and globular of Figure 2) illustrated by Habib (1979*a*, pl. 2). In the shredded xenomorphic facies, morphologically sorted pollen grains range from as many as 40 000 per gram to as few as 500 or less. Tracheids are largely absent or are represented by carbonized fragments, and the dinoflagellates are well-represented. The palynological origin of the shredded debris is not known, but it may represent organic detritus of various origins degraded by bacterial activity at the site of burial during periods of diminished rates of sedimentation. It occurs in all sections in intervals where there is evidence of lower rates of sedimentation (e.g. Site 99a; Site 387, Tucholke, Vogt *et al.*, 1979). The micrinitic facies represents the farthest offshore site of terrigenous organic sedimentation, in black clays. It consists of abundant black-colored, carbonized debris representing the much altered woody tissue of land plants which became resistant to further chemical or bacterial oxidation. Pollen grains and

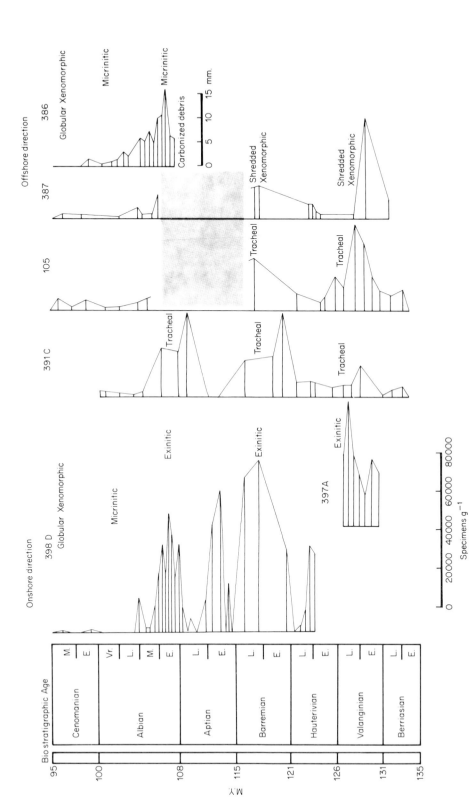

Figure 3. Hypothetical trend away from continental margins showing progressive modification of terrigenous organic matter in the exinitic, tracheal, shredded xenomorphic, and micrinitic facies during episodes of increased sedimentary supply. Curves of Sites 398D-387 represent abundances of structured terrigenous matter in specimens per gram of sediment. Site 386 curve is the amount of terrigenous carbonized debris measured in a 2-dram vial, in the absence of structured matter.

dinoflagellates are few and generally poorly preserved, consisting of a larger number of species of dinoflagellates than of the morphologically sorted pollen grains.

The globular xenomorphic facies differs from the previously described facies in the nature of the amorphous debris and the very few but well-preserved pollen grains and dinoflagellates. Pyrolysis assay of the intervals containing this facies indicates that the organic carbon is of marine origin (Tissot *et al.*, 1979).

Episodes of increased supply of terrigenous organic matter are essentially contemporaneous in the North Atlantic, as exemplified in the sections from Sites 398D and 391C. These range from Late Hauterivian to Early Aptian and from Late Aptian to early Middle Albian, respectively. The Albian pulse is not represented at Sites 105 and 387 (Figure 3), due presumably to a biostratigraphic disconformity determined from the truncation of dinoflagellate stratigraphic ranges in these sections (Habib, 1977). Figure 4 illustrates the graphic-expression-of-correlation

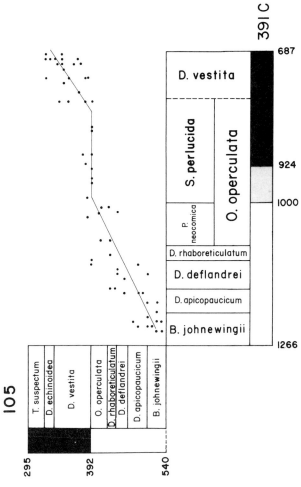

Figure 4.   Shaw's graphic-expression-of-correlation technique comparing ranges of mutually occurring species at Sites 105 and 391C. Horizontal plateau in line of correlation is suggestive of one or more disconformities at the approximate contact between lower calcareous sediments and upper black clays. Note the disproportionately thinner *Odontochitina operculata* zone at Site 105, indicating a missing interval which involves at least part of the late Barremian-early Albian at Site 391C. Dinoflagellate zonation after Habib (1977).

technique formulated by Shaw (1964) where, by plotting the lowermost and uppermost occurrences of species common to two sections, a plateau in the line of correlation is indicative of a missing stratigraphic interval in one of them. Relative to Site 391C, the missing interval at Site 105 involves the equivalent of at least part of Aptian calcilutites and Early Albian black clays. The disconformity is expressed best where there is a relatively sharp contact between earlier Cretaceous nannofossil limestones and middle Cretaceous black clays. It is widespread in the western North Atlantic, occurring also at or near this lithologic contact at Sites 387 and 101A (Figure 1). The stratigraphic position of the disconformity, at the base of the black clays, corresponds to the position of seismic Horizon Beta, widespread in the western North Atlantic (Ewing & Hollister, 1972).

A third injection of terrigenous organic matter is evident in the Valanginian of sites 397A–387 (Figure 3). At Site 397A, 150 m of Cretaceous prodeltaic laminated muds (von Rad, Ryan et al., 1979) contains a well-developed exinitic facies (Figure 5). The middle part, represented by cores 41 through 46, contains a number of Neocomian dinoflagellate species. Although rare in the samples, the presence of *Druggidium apicopaucicum*, *Biorbifera johnewingii*, and *Diacanthum hollisteri* is indicative of a Valanginian age, which is in agreement with the nannofossil age assigned to this section by H. Thierstein (in von Rad, Ryan et al., 1979). The large amount of organic matter in the exinitic facies, especially in the upper part of the section, is presented as evidence that the rate of supply of terrigenous matter was higher during this corresponding period of time.

Trends away from the continental margins during episodes of increased terrigenous supply (Figure 3) illustrate the morphological sorting of pollen grains and the reduction in numbers of specimens of the structured material (pollen grains, spores, tracheids) proposed in the sedimentary model. The exinitic facies is poorly sorted, palynologically, in sections nearest the margins (Figure 1). Farther from the sources of the major influx of land plant materials, the three episodes show progressive sedimentological modification in the sorting out of large fern spores, concentration of numerous tracheids and pollen grains morphologically adapted for transportation and admixing of dinoflagellates (in the tracheal facies), reduction in the number of the sorted pollen grains and the loss of large tracheids along with the

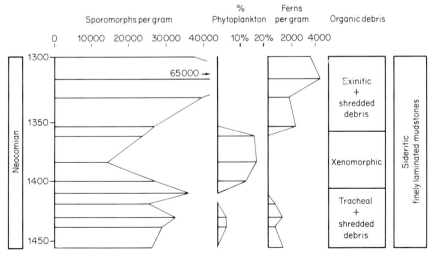

Figure 5.   Distribution of organic facies at Site 397A. Dinoflagellate cysts in the shredded xenomorphic interval indicate a Valanginian age.

substantial increase of shredded debris and dinoflagellates (shredded xenomorphic facies), and finally very few to no pollen grains (or dinoflagellates) in the concentration of very small carbonized debris (micrinitic facies). In the Late Hauterivian–Aptian interval, the total amount of all structured terrigenous material is reduced from an average of approximately 55 000 specimens per gram in the Site 398D section to approximately 37 000 at Sites 105 and 391C to a low of approximately 9 000 at Site 387. At Site 386, farthest removed from the continental margins, the early Middle Albian terrigenous event is expressed entirely by the increased amount of carbonized comminuted woody tissue (micrinitic facies) in the almost complete absence of pollen grains and dinoflagellate cysts (Tucholke, Vogt et al., 1979, p. 227). This corresponds well with the interpretation that this carbonized woody tissue was resistant to further degradation and could be carried into the farthest offshore areas of an oxic ocean (Cross et al., 1966; Tissot et al., 1979).

The micrinitic facies of Middle Albian–Early Cenomanian age is distributed throughout Sites 398D–386 (Figure 3). The occurrence of abundant carbonized debris and few pollen grains, most of which are morphologically sorted and poorly preserved, in sections ranging from close to the continental margins to as far as the central part of the North Atlantic reflect their increased distance from shorelines (and source of land plant materials) migrating onto the continents during the world-wide marine transgression which commenced in the Albian. Thus the transgressive Albian–Cenomanian terrigenous black clays at Site 398D may have been deposited in a sedimentary locus relative to the shoreline similar to that of Sites 387 and 386 during the earlier Albian deltaic terrigenous event. The number of pollen grains in the micrinitic facies is small in all the investigated sections. Nevertheless, there is the same reduction in supply progressively offshore, from an average maximum of approximately 5000 specimens per gram at Sites 398D, 391C and 105 to less than 500 at Site 387 to almost none at 386.

The organic facies characterized by abundant globular amorphous debris is also geographically extensive, in those sections containing sediments as young as Middle or Late Cenomanian. Pollen grains and dinoflagellates are well-preserved but few in number and, in contrast to the largely terrigenous matter deposited during the Early Cretaceous, formed in an environment which accumulated abundant marine organic carbon, according to pyrolysis assay (Tissot et al., 1979). Palynology cannot identify the biological or sedimentary source of this debris.

The globular xenomorphic facies was formed at the time of the maximum marine transgression of early Late Cretaceous age. Terrigenous organic matter is minimal due to the locus of the sections farthest removed from transgressing shorelines. Dinoflagellate fossils are not numerous, since the fossilizable cyst-producing species were restricted to paleoshelf environments distant from the investigated sections, although they became increasingly diversified through the Middle Albian–Cenomanian (Habib, 1979a). The expansion of the neritic facies, brought about by the marked increased area of continental shelves submerged during the maximum transgression, is considered to be responsible for a significant increase in marine organic productivity on the shelves (Schlanger & Jenkyns, 1976; Jenkyns, 1980) and perhaps also through the food chain and increased supply of nutrients in the planktonic oceanic environment. For example, evidence of sea surface fertility which produced rich organic matter in the open ocean is found in several thin layered black mudstones containing abundant radiolarian fossils in the Cenomanian at Site 398D (Ryan, Sibuet et al., 1979). It was during this period of time that cyst-producing dinoflagellates reached a pronounced peak of diversification (Tappan & Loeblich, 1971) in the marginal and epeiric seas, with the introduction of at least 150 new species through the Albian (Bujak & Williams, 1979). The transgression is related to the major radiation of benthic invertebrate fauna based on the sharply increased diversification of bivalve genera (Kauffman, 1973, especially Figures 6 and 7).

The weight-percentage distribution of organic carbon shows a generally consistent correspondence with the distribution of the palynologically defined organic facies, and is considered to have varied as the consequence of the same sedimentary processes. Except for the Cenomanian marine-carbon event, the highest percentages of organic carbon occur in intervals of Early Cretaceous age containing the most abundant terrigenous organic matter (Deroo *et al.*, 1978; Habib, 1979*a*, 1979*b*), in lithologies ranging from turbiditic siltstones and fine sandstones to laminated limestones to black clays. Increased supply of land-plant materials during the terrigenous episodes correlates with enriched organic carbon of almost exclusively terrestrial origin, in sections close to a deltaic source (Site 398D) and farthest offshore (Site 386). In those terrigenous intervals where the rate of sedimentation could be reliably calculated, e.g. Site 398D (Ryan, Sibuet *et al.*, 1979), high rates of terrigenous mineral sedimentation correlate with increased organic carbon percentages as well. Organic carbon percentages are *relatively* higher within the terrigenous intervals of each section despite the offshore sorting of organic matter. Thus, the percentages of terrestrial carbon are highest in the poorly sorted exinitic facies and are also relatively higher in the terrigenous interval at Site 386, due to the increased amounts of carbonized debris in that part of the section (Figure 6).

The interval of much diminished supply of terrigenous organic matter correlates with a much smaller terrestrial carbon content and with lower rates of sedimentation, in the transgressive phase which deposited the widespread micrinitic facies of the Middle Albian–Early Cenomanian. This facies consists largely of the blackish fine fragments of carbonized woody tissue which are considered to be largely responsible for the black color of the organic carbon-poor clays in which they are concentrated.

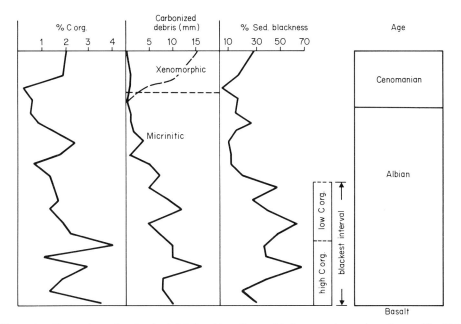

Figure 6.   Comparison of carbonized debris with sediment blackness and organic carbon at Site 386. Carbonized debris remains higher in blackest interval despite lower percentages of organic carbon. Otherwise the three curves are quite similar. Dashed curve represents increase of globular amorphous debris in Middle/Late Cenomanian. Organic carbon and sediment blackness curves after McCave (1979).

McCave (1979) attributed the blackish color of the organic carbon-poor shales at Site 386 to the concentration of finely disseminated pyrite in the darkest colored interval although Curtis (1980) and Berner (1974) state that the color of present-day black muds (in the Black Sea for example) is due to iron monosulfides which are not found in ancient black shales. However, these monosulfides generally mature into pyrite with time. Figure 6 shows that the distribution of carbonized debris closely follows the sediment–blackness curve, which suggests that it is the color of the debris, and its larger amounts, which contributed significantly to the color of the sediment. According to palynological interpretation, the increased amount of carbonized debris in the earlier Albian at Site 386 represents a relatively higher rate of terrigenous sedimentation than in the younger Albian sediments of that section. The stratigraphically lower interval is synchronous with the exinitic facies deposited much closer to a land plant source and at a much more rapid rate (Figure 3).

Site 387 (Figure 7) is representative of an offshore site which ranges through the investigated chronostratigraphic sequence; it shows the range of organic facies through the Lower Cretaceous–Cenomanian. On biostratigraphic evidence, a large part of the Aptian–Albian is missing, although abundant carbonized debris in black clays immediately above the disconformity may be synchronous with the earlier Albian interval at Site 386. At Site 387, the terrigenous episodes of Valanginian and Barremian–Aptian ages correlate with organic carbon content, although the Valanginian interval is not as well established. Nevertheless, both terrigenous intervals are synchronous with those in the other investigated sections. The offshore

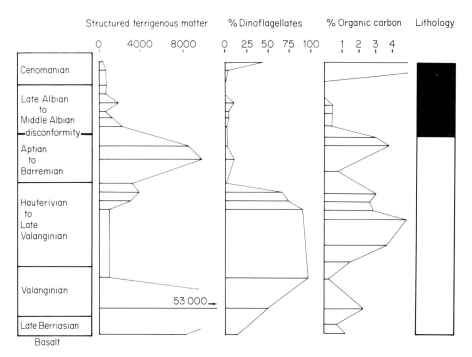

Figure 7. Comparison of distribution of organic matter and organic carbon at Site 387. Barremian–Aptian and Valanginian episodes of increased terrigenous organic matter correlate with increased organic carbon. Aptian–Albian terrigenous event lost presumably to disconformity near lithologic contact, as inferred from truncation of dinoflagellate and pollen stratigraphic ranges. Increased organic carbon in the Hauterivian due to abundance of dinoflagellate systs and is interpreted to represent increased marine carbon. Increased organic carbon and dinoflagellates occur in the Hauterivian at Sites 398D and 105 as well (Habib, 1979a, Figures 3 and 4).

character of the intervals is well-expressed by the abundance of shredded amorphous debris, fewer and well-sorted pollen grains, and the occurrence of few and largely carbonized tracheids.

Organic carbon is especially rich in the Cenomanian marine carbon event at Site 387. Similar to the other sections, terrigenous organic matter is very poorly represented in an organic facies rich in globular debris. However, there is another interval of higher organic carbon, in the Hauterivian Stage (Figure 7). This interval is especially rich in dinoflagellate cysts and very poor in terrigenous organic matter. This palynological evidence requires a significant admixing of marine carbon during this episode of organic sedimentation. The larger amounts of dinoflagellates generally in the tracheal and shredded xenomorphic facies is cited as evidence of a proportional increased contribution of marine carbon in a general offshore direction.

3. Origin of Cretaceous black shales

Sedimentary processes acting on and near the outer margins of the continents bordering the Cretaceous North Atlantic were responsible for contributing large amounts of organic matter of both terrestrial and marine origins. Geographically extensive Early Cretaceous deltas prograding at regressing shorelines very close to the present-day outer limits of continental shelves discharged large amounts of terrigenous organic matter into the ocean which, by redistribution towards the center of the North Atlantic, controlled the organic carbon character of both organic-rich and organic-poor sediment, as well as its color. During the Middle Cretaceous transgressive phase, the offshore organic-poor shales were deposited closer to the present continental margins in response to the continentward migration of shore-lines. The marked increase of marine organic productivity, exemplified by sharply increased diversification of dinoflagellates and bivalves in the marginal seas and presumably to that of the planktonic fauna and flora in the open ocean, supplied abundant organic carbon to the organic-rich and widespread sediments of Cenomanian age.

The central tenet of the sedimentary-supply hypothesis is that the rate of burial, related to the locus of deposition of terrigenous materials and to productivity, governed the nature and amount of organic matter. Curtis (1980) stressed the importance of the extent and style of bacterial diagenetic activity in the uppermost few meters of buried sediment as a function of the rate of burial. Sufficiently high rates of organic sedimentation would transport organic matter quickly through the aerobic bacterial oxidation zone. Lower sedimentation rates would expose organic matter in the aerobic zone. According to Curtis (1980, citing Love, 1967 for an example), muds containing abundant organic matter rapidly become anoxic even under fully oxygenated bottom waters, and dark-colored sediments rich in organic matter are commonly found today beneath fully oxygenated marine waters and in the intertidal zone.

In the North Atlantic, the amount of organic matter supplied and its rate of burial were the dominant factors controlling the distribution of organic carbon. During high rates of organic sedimentation, as in the exinitic and tracheal facies in prodeltaic and other turbiditic terrigenous intervals, organic matter was buried rapidly and was relatively unaltered by diagenetic oxidative processes. Evidence for this conclusion is taken from the excellent preservation of numerous and diverse materials, e.g. pollen grains, spores, dinoflagellates, acritarchs, tracheids, cuticles, colonial algae, and megaspores. Lower rates of sedimentation in progressively offshore terrigenous intervals (shredded xenomorphic and micrinitic facies) deposited lesser amounts of these materials and increased amounts of altered amorphous debris in organic-poor sediments. The composition of the micrinitic facies, consisting of poorly-preserved pollen grains and carbonized woody tissue, is presented as evidence of lower rates of terrigenous sedimentation and longer exposure in the aerobic oxidation zone in the farthest offshore setting during both the Early Cretaceous deltaic phases as well as in the Middle Albian transgressive. The color of these organic carbon-poor black shales is not the result of the amount of organic carbon *per se*, but primarily of the color of the carbonized debris.

The globular facies in the organic-rich shales of Cenomanian age may have accumulated through small periods of time at a rapid rate because of organic productivity, even in the absence of rapid mineral sedimentation. Pre-existing ocean-wide anoxicity is an unnecessary alternative, when one considers that the Cenomanian North Atlantic already was a large ocean, areally more extensive than ever before in its Mesozoic history, because of transgressing shorelines and the stage of seafloor spreading (Kauffman, 1973).

The origin of globular amorphous debris remains a mystery. It is the most abundant constituent of the facies it defines, and occurs together with few, very well-preserved, dinoflagellates and pollen grains in an interval of organic carbon geochemically analyzed to be of marine origin. This debris may well be the remains of marine organisms such as the open-ocean planktonic dinoflagellates which were not fossilized. The cyst-forming (and thus fossilizable) species of dinoflagellates form only a minor part of the extant dinoflagellate flora. If the proportions in the Cretaceous ocean were at all similar to those in the modern oceans, then the sharp increase in the cyst-producing species during the middle Cretaceous would have been matched by a proportional and therefore much greater diversification of the open ocean nonfossilizing flora. An indirect measure of this inferred dinoflagellate productivity may be found in the geochemical analysis of the globular debris specifically, which should indicate if this material is actually marine in origin, and in the analysis of phosphate content. Phosphorous is an element which is considered to reflect the magnitude of sea-surface productivity (Ryan & Cita, 1977). Fauconnier & Slansky (1978) noted the relatively high content of phosphorous in the mineral fraction of modern dinoflagellates and the abundance of this ocean plankton close to modern and subrecent phosphate nodule deposits. In their study of the Albian–Cenomanian neritic facies of the Paris Basin, they showed a similarity in the sequences of numerous cysts per gram of sediment (and paucity of terrigenous organic matter) and phosphate abundance, and suggested a relationship between the two. If the globular debris is the result of increased marine organic productivity, including increased dinoflagellate productivity, then the black shales containing this facies may well have become phosphate-enriched. For example, Arthur and Premoli Silva (1980) described a phosphatic organic-rich radiolarian mudstone of Late Cenomanian–Early Turonian age from the Alpine Tethys.

In conclusion, it is not necessary to formulate models which require conditions to preserve organic matter regardless of its sedimentary supply, at least for the investigated North Atlantic Cretaceous. The bottom water may have been anoxic or oxic, but did not seriously affect the organic carbon content because of the abundance of organic matter supplied and of its sufficiently rapid burial. Where the organic matter was supplied at a rapid rate (exinitic facies), it may actually have caused anoxic conditions in the immediately overlying waters (Gardner *et al.*, 1978); where it accumulated more slowly (micrinitic facies), it appears that the bottom waters would have to have been oxidizing for the diagenetic aerobic oxidation of organic matter in the early burial environment to work (Curtis, 1980).

The effects of the organic events are of a large scale. Episodes of organic sedimentation were penecontemporaneous across the North Atlantic (Habib, 1979*a*) in response to the deltaic and then transgressive processes which simultaneously affected the continental margins bordering this ocean. The timing of the Aptian–Albian and Cenomanian black shale events corresponds to that of the global 'oceanic anoxic events' described by Schlanger & Jenkyns (1976). The widespread distribution of these blackish sediments in various paleo-oceanographic and paleobathymetric settings requires that different processes were responsible for black shale formation at different places. If these global black shale events were the consequence of a single overriding factor, then it was most likely an ultimate response of the increase of both the terrestrial and marine biota to changes in atmospheric climate. The Early Cretaceous was a time of humid climate compared with the more arid climate of the Jurassic, and was equable compared with the climatically steepened gradients of the Cenozoic. Deltas discharged large volumes of fresh water, rich in both organic detritus and dissolved nutrients, into oceans and thus probably contained much greater amounts of organic matter than the present North Atlantic does for example. Major deltaic activity implies increased continental runoff due to increased atmospheric precipitation. The palynology of Cretaceous sediments attests to the rise in atmospheric moisture; schizaeaceous fern spores diversify beginning near the Jurassic–Cretaceous boundary (Norris, 1969) and angiosperm pollen diversify rapidly in the Barremian–Cenomanian (Doyle, 1969).

Further study of black shale deposits may show that the Cretaceous "anoxic events" were not unique. Black shale formation may not have been so uncommon in ocean basin history.

Acknowledgments

My appreciation is extended to Jonathan Bujak of the Bedford Institute of Oceanography (Canada) for permission to cite unpublished information on the C.O.S.T. B-2 well, and to Richard Olsson of Rutgers University, New Jersey, USA for providing me with samples from it. Geoffrey Eaton (British Petroleum Ltd.) and David Locke (Queens College Chemistry Dept., USA) reviewed the manuscript. Appreciation is extended to participants on board D.V. *Glomar Challenger* Leg 76 drilling in the Blake–Bahama Basin for fruitful discussion on this subject. M. Zotto and R. S. Habib assisted in the microscope work and preparation of samples. C. Robertson prepared the diagrams. M. Rafanelli typed the revised manuscript. This study was supported by the U.S. National Science Foundation, grant OCE-79-13191 at Queens College.

References

Allen, P. 1969. Lower Cretaceous sourcelands and the North Atlantic. *Nature* **222**, 657–658.

Arthur, M. A. & Natland, J. H. 1979. Carbonaceous sediments in the North and South Atlantic: the role of salinity in stable stratification of early Cretaceous basins. *Deep Drilling Results in the Atlantic Oceans Continental Margins and Paleoenvironment. Maurice Ewing Series 3, Am. Geophys. Union* (Eds M. Talwani, W. Hay & W. B. F. Ryan), 375–401.

Arthur, M. A. & Premoli Silva, S. 1980. Developments of widespread carbonaceous sediments in the Cretaceous Tethys. *26th International Geological Congress Abstracts* 198.

Batten, D. 1973. Use of paynologic assemblage types in Wealden correlation. *Palaeontology* **16**, 1–40.

Benson, W. E., Sheridan, R. E. *et al.* 1978. *Initial Reports of the Deep Sea Drilling Project* **44**.

Berner, R. A. 1971. *Principles of Chemical Sedimentology*. McGraw-Hill, New York, pp. 1–240.

Berner, R. A. 1974. Iron sulfides in Black Sea sediments and their paleo-oceanographic significance. *The Black Sea: Its Geology, Chemistry and Biology* (Eds E. T. Degens & D. A. Ross) *A.A.P.G. Mem. 20*, 524–531.

Bujak, J. P. & Williams, G. L. 1979. Dinoflagellate diversity through time. *Marine Micropaleontology* **4**, 1–12.

Cross, A. T., Thompson, G. G. & Zaitzeff, J. B. 1966. Source and distribution of palynomorphs in bottom sediments southern part of Gulf of California. *Marine Geology* **4**, 467–524.

Curtis, C. D. 1980. Diagenetic alteration in black shales. *Journal of the Geological Society of London* **137**, 189–194.

Degens, E. T. & Ross, D. A. (Eds). 1974. The Black Sea: Its Geology, Chemistry and Biology. *Mem. American Assoc. Petroleum Geologists* **20**, 1–663.

Deroo, G., de Graciansky, P. C., Habib, D. & Herbin, J.-P. 1978. L'origine de la matiere organique dans les sediments cretaces du site I.P.O.D. 398 (haut-fond de Vigo): correlations entre les donnes de la sedimentologie, de la geochimie organique et de la palynologie. *Bulletin de la Societé geologique de France* **20**, 465–469.

Doyle, J. A. 1969. Cretaceous angiosperm pollen of the Atlantic Coastal Plain and its evolutionary significance. *Jour. Arnold Arboretum* **50**, 1–35.

Einsele, G. & von Rad, U. 1979. Facies and paleoenvironment of Lower Cretaceous sediments at DSDP site 397 and in the Aaiun basin (northwest Africa). *Initial Reports of the Deep Sea Drilling Project* **47** (1), 559–577.

Ewing, J. I. & Hollister, C. D. 1972. Regional aspects of deep sea drilling in the western North Atlantic. *Initial Reports of the Deep Sea Drilling Project* **11**, 951–976.

Fauconnier, D. & Slansky, M. 1978. Role possible des dinoflagelles dans la sedimentation phosphatee. *Bull. Bureau Recherches Geologiques et Minieres* **4**, 191–200.

Fischer, A. G. & Arthur, M. A. 1976. Secular variations in the pelagic realms. *Deep Water Carbonates and Environments* (Eds H. E. Cook & P. Enos). *Soc. Econ. Paleontol. Mineral. Spec. Publ.* **25**, 19–50.

Gardner, J. V., Dean, W. E. & Jansa, L. 1978. Sediments recovered from the northwest African continental margin. *Initial Reports of the Deep Sea Drilling Project* **41**, 1121–1134.

Gignoux, M. 1955. *Stratigraphic Geology*. W. H. Freeman and Co. 1–682.

Habib, D. 1968. Spores, pollen, and microplankton in deep sea cores from the Horizon Beta outcrop. *Science* **162**, 1480–1481.

Habib, D. 1977. Comparison of Lower and Middle Cretaceous palynostratigraphic zonations in the western North Atlantic. *Stratigraphic Micropaleontology of Atlantic Basin and Borderlands* (Ed. F. M. Swain). Elsevier Publishing Company. Developments in Paleontology and Stratigraphy, **6**, 341–367.

Habib, D. 1979a. Sedimentary origin of North Atlantic Cretaceous palynofacies. *Deep Drilling Results in the Atlantic Ocean: Continental Margins and Paleoenvironment* (Eds M. Talwani, W. Hay & W. B. F. Ryan). *Maurice Ewing Series 3, American Geophysical Union*, 420–437.

Habib, D. 1979b. Sedimentology of palynomorphs and palynodebris in Cretaceous carbonaceous facies south of Vigo Seamount. *Initial Reports of the Deep Sea Drilling Project* **47** (2), 451–465.

Jansa, L. & Wade, J. A. 1975. Geology of the continental margin off Nova Scotia and Newfoundland. *Geological Survey of Canada. Paper 74–30* **2**, 51–105.

Jenkyns, H. C. 1980. Cretaceous anoxic events: from continents to oceans. *Journal of the Geological Society of London* **137**, 171–188.

Kauffman, E. G. 1973. Cretaceous bivalvia. In *Atlas of Paleobiogeography* (Ed. A. Hallam). Elsevier Publishing Company, 353–384.

Love, L. G. 1967. Early diagenetic iron sulfide in recent sediments of the Wash (England). *Sedimentology* **9**, 327–352.

McCave, I. N. 1979. Depositional features of organic carbon-rich black and green mudstones at DSDP sites 386 and 387, western North Atlantic. *Initial Reports of the Deep Sea Drilling Project* **43**, 411–416.

Minard, J. P., Perry, W. J., Weed, E. G. A., Rhodehamel, E. C., Robbins, E. I. & Mixon, R. B. 1974. Preliminary report on geology along Atlantic continental margin of northeastern United States. *Bulletin of the American Association of Petroleum Geologists* **58**, 1169–1178.

Norris, G. 1969. Miospores from the Purbeck beds and marine Upper Jurassic of southern England. *Palaeontology* **12**, 574–620.

Ryan, W. B. F. & Cita, M.-B. 1977. Ignorance concerning episodes of ocean-wide stagnation. *Marine Geology* **23**, 197–215.

Ryan, W. B. F., Sibuet, et al. 1979. Site 398. *Initial Reports of the Deep Sea Drilling Project* **47** (2), 25–233.

Schlanger, S. O. & Jenkyns, H. C. 1976. Cretaceous oceanic anoxic sediments: causes and consequences. *Geologie en Mijnbouw* **55**, 179–184.

Shaw, A. B. 1964. *Time in Stratigraphy*. McGraw-Hill, New York, 1–365.

Tappan, H. & Loeblich, A. R. Jr. 1971. Geobiologic implications of fossil phytoplankton and time-space distribution. In *Palynology of the Late Cretaceous and Early Tertiary* (Eds B. M. Kosanke & A. T. Cross). *Geological Society of America. Special Paper* **127**, 247–340.

Thierstein, H. R. & Berger, W. H. 1978. Injection events in ocean history. *Nature* **276**, 461–466.

Tissot, B., Deroo, G. & Herbin, J. P. 1979. Organic matter in Cretaceous sediments of the North Atlantic: contribution to sedimentology and paleogeography. *Deep Drilling Results in the Atlantic Ocean: Continental Margins and Paleoenvironment. Maurice Ewing Series 3* (Eds M. Talwani, W. Hay & W. B. F. Ryan), *American Geophysical Union*, 362–374.

Tucholke, B. E., Vogt, P. R. *et al.*1979. *Initial Reports of the Deep Sea Drilling Project* **43**.
von Rad, U., Ryan, W. B. F. *et al.* 1979. *Initial Reports of the Deep Sea Drilling Project* **47**.
Windisch, C. C., Leyden, R. J. Worzel, J. L., Saito, T. & Ewing, J. 1968. Investigations of Horizon Beta.
    Science **162**, 1473–1479.

Note added in proof

Subsequent to this study, Habib (1982) identified this amorphous debris in the form and size
of zooplanktonic fecal pellets, very close to that produced by the modern copepod *Cyclops
sentifer* fed a diet of dinoflagellates and other phytoplankton (Porter & Robbins, 1981).

Habib, D. 1982. Sedimentation of black clay organic facies in a Mesozoic oxic North Atlantic. *Third
    North American Paleontological Convention* **1**, 217–220.
Porter, K. G. & Robbins, E. I. 1981. Zooplankton fecal pellets link fossil fuel and phosphate deposits.
    Science **212**, 931–933.

7. Some Remarks about the Stable Isotope Composition of Cyclic Pelagic Sediments from the Cretaceous in the Apennines (Italy)

P. L. de Boer

Comparative Sedimentology Division, Institute of Earth Sciences, Utrecht

The Middle Cretaceous pelagic sequence in the Apennines is characterized by cyclic lithological changes, mostly represented by an alternation of carbonate-rich and marly beds. Stable oxygen isotope data point to a causal relationship with regular changes of temperature. It is suggested that these are caused by regular fluctuations of the velocity of ocean water circulation in response to shifts of the caloric equator due to astronomical influences.

Factors which may have influenced the stable carbon isotope composition in the sequence studied are discussed. (1) The stable carbon isotope ratio of the organic matter seems to be related to the relative amount of terrestial input. (2) Within black shales, it is suggested that bacterial fermentation may have caused an increase of $\delta^{13}C$ of the carbonate. (3) The $\delta^{13}C$ values of carbonate in sequences which were deposited under continuously oxygenated conditions seem to be related to the rate of productivity and the removal of organic matter by sinking into deep water.

1. Introduction

Large amounts of organic carbon were stored in pelagic sediments, especially within the Tethyan and North Atlantic domain during the Middle Cretaceous (Schlanger & Jenkyns, 1976; Fischer & Arthur, 1977; Ryan & Cita, 1977; Thiede & van Andel, 1977; Arthur, 1979; Jenkyns, 1980). Many authors explain this observational fact by assuming that in this epoch the great rise of sea-level (Hays & Pitman, 1973) caused a large global transgression, and that ocean waters were relatively warm (Frakes, 1979), and had a low circulation velocity (Berger, 1979), thereby introducing climatological and oceanographic models which explain the exhaustion of oxygen in deep water.

The observed shift of the carbon isotope composition of carbonates of that age to relatively positive values reflects the extraction of large amounts of organic matter ($\delta^{13}C \sim -25\%_{oo}$) from the Cretaceous oceanic carbon reservoir (Scholle & Arthur, 1980; Veizer, Holser & Wilgers, 1980).

Weissert, McKenzie & Hochuli (1979) described black shales, organic carbon-rich marly and clayey pelagic sediments, of Lower Cretaceous age in the Southern Alps. These were deposited continuously, but in regular alternation with sediments containing less or no organic matter. The authors suggest that the relatively positive $\delta^{13}C$ values of carbonate in the black shale intervals are the result of an increase of $\delta^{13}C$ of HCO_3^{-} in the surface water, caused by the removal of organic matter into the sediment during deposition of black shales, analogous to the effect of long-term distraction of organic carbon. A critical evaluation of this statement is part of the subject of this paper.

Rhythmic alternations of lithology are also found in places where no anoxic conditions occurred during deposition. Often such a rhythmicity is defined by a regular variation of the carbonate content within the sedimentary succession. Such a succession is discussed in this paper; the outcrop area is the Northern Apennines where the sediments were deposited in a pelagic environment with a sedimentation rate of 1 cm per 1000 years or less (Wonders, 1980). Their age is Late Barremian to Turonian.

The rhythmicity observed was related to astronomical parameters, i.e. the precession cycle (de Boer & Wonders, 1981). It was suggested that the link that exists between astronomical variables and the rhythmicity in such pelagic sequences is formed by an astronomically defined shifting of the tropical upwelling and high productivity zone (de Boer, 1982).

In order to further verify this assumption, and to test the usefulness of detailed studies of the stable carbon isotope ratio of pelagic carbonates, the bank to bank variations of the stable isotope composition of carbonate and some samples of organic matter were analysed.

2. Material

The sections from which samples were analysed for stable isotope composition, are located in the Umbrian Apennines, Italy. The first (Figure 1) is from Upper Barremian strata west of Acqualagna (43° 39′N, 12° 43′E) and is characterized by an alternation of white carbonate-rich intervals and black marls, rich in organic matter and lacking traces of burrowing or benthonic life. The larger part of the carbonate consists of calcareous nannofossils. The content of organic matter in the black shale intervals is 1% or more; carbonate content ranges from less than 40% in black shale intervals to 90% in the carbonate beds.

The second section is from the "Marne a Fucoidi" (Aptian–Albian) near Moria (43° 33′N, 12° 41′E; Figure 2).

○ carbonate

● black shale

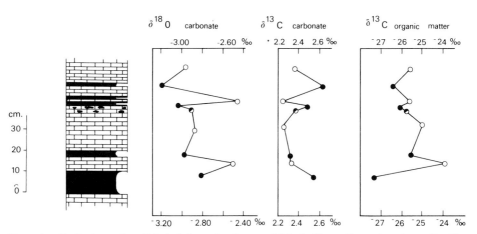

Figure 1.   Cyclic alternation of black shales and carbonate-rich beds; Upper Barremian, river valley 6 km west of Acqualagna, Northern Apennines.

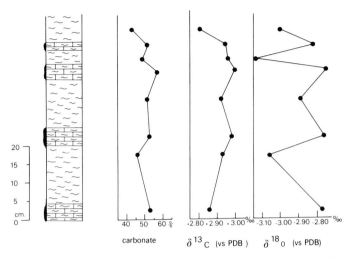

Figure 2.   Marne a Fucoidi, Aptian/Albian, Moria, Northern Apennines.

In the field, a rhythmic banding is visible, due to an alternation of reddish and greenish beds. The reddish beds are slightly more erosion resistant than the green ones, and have a higher carbonate content (Figure 3).

The third studied interval (Figures 4, 5; same coordinates as Figure 2) is from the Albian part of the "Scaglia Bianca" near Moria. This sequence consists of a regular alternation of carbonate-rich and carbonate-poorer beds or marls. Locally, the marly parts show a dark-grey to black colour, which here seems to be due to the presence of pyrite rather than of organic matter; the maximum amount of organic matter found within the dark-coloured intervals is 0.66% (van Graas, Viets, de Leeuw & Schenck, 1981).

Figure 3.   Marne a Fucoidi; on the left-hand side is the section described in Figure 2.

The contrast between "carbonate-rich" and "carbonate-poor" or marly intervals, as seen in the different outcrops, seems to be the result of weathering, rather than of large variations of carbonate content. Differences of $CaCO_3$ content between adjacent intervals, as low as 5–10%, may already lead to a differential weathering which makes the carbonate-poorer parts of the sequence look much softer than adjacent "carbonate-rich" intervals in the field. In fresh exposures, created by the removal of some decimeters of sediment, the difference is less clear or not visible at all.

The terms "carbonate, carbonate-rich", and "marly, carbonate-poor", used below, refer to the content of carbonate relative to that of adjacent beds, rather than to the absolute carbonate content. In the same way, terms like "high productive" and "low productive" are relative notions; as explained below, the surface waters which supplied the biogenic part of the sediments, are assumed to have been, in an absolute sense, low productive continuously.

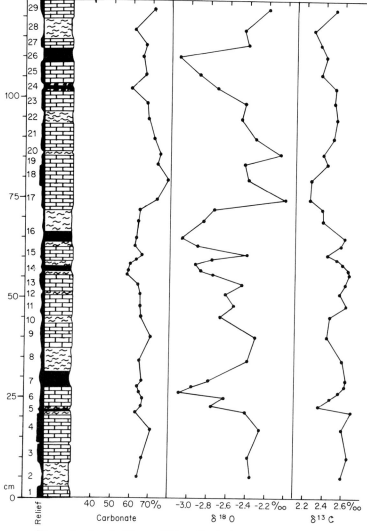

Figure 4.   Scaglia Bianca, Albian, Moria, Northern Apennines.

Figure 5. Section described in Figure 4; the numbers refer to the numbering of beds in Figure 4.

3. Analytical method and results

Samples containing much organic carbon were heated for 30 minutes at 470°C in a helium flow, in order to deactivate the organic matter. After reaction of the powdered samples with concentrated H_3PO_4 at 25°C under initial vacuum conditions for at least four hours, the liberated CO_2 was led through a cold trap (melting aceton, −96°C), in order to remove H_2O, and then was trapped by means of liquid nitrogen. For the analyses of the organic carbon, carbonate was removed by treatment with diluted HCl; after neutralizing and drying, samples were heated at 900°C in a 0.2 O_2 atmosphere for 20 minutes. The oxidation products were trapped by means of liquid nitrogen, which was next replaced by melting aceton in order to allow CO_2 to evaporate and to retain water. Before measuring, CO_2 was led through a cold trap once again to remove the last traces of H_2O.

The samples were analysed with a Micromass 602 c mass spectrometer. Isotope data are relative to PDB standard. Duplicates generally do not differ more than 0.05‰. A systematic difference between duplicates which were analysed some months after each other, is ascribed to a gradual change of the standard gas, and has been taken into account.

In general, the carbonate in the carbonate-rich beds contains relatively heavy oxygen.

The pattern of the carbon isotope composition of carbonate is not similar in the various sections. In the one with true black shales (Figure 1), the lower $\delta^{13}C$ values are found within the carbonate-rich beds. In the Moria section of Figure 2, the $\delta^{13}C$ values of carbonate in the carbonate-richer parts show a slightly positive deviation from the mean trend. The pattern of $\delta^{13}C$ values in Figure 4 is not very distinct. The $\delta^{13}C$ values of black intervals are relatively high compared to the values of carbonate-rich beds. In light-coloured marly beds, lower values are found, but the differences are very small and not very significant. In the Barremian section (Figure 1) the carbon isotope composition of the bulk organic matter shows a trend which is contrary to the trend of the carbon isotope composition of the carbonate in the same samples.

4. Stable oxygen isotopes

The variation in stable oxygen isotope values measured on carbonate in successive beds tends to follow the fluctuation of carbonate content: beds with relatively high carbonate content show higher $\delta^{18}O$ values and vice versa. An explanation has to be sought in either primary or secondary processes, or in a combination of both.

Diagenetic processes may lead to changes of $\delta^{18}O$ values of the carbonate. Arthur (1977) shows that carbonate-rich beds can be enriched in $CaCO_3$ during diagenesis, at the expense of marly intervals. Precipitation of carbonate during diagenesis generally results in a lowering of the $^{18}O/^{16}O$ ratio of bulk carbonate (cf. McKenzie, Bernouilli & Garrison, 1978). Therefore, precipitation of $CaCO_3$ within carbonate-rich intervals is likely to have resulted in a lowering of $\delta^{18}O$ values compared to the original values. If the rhythmic carbonate–marl successions were uniquely the result of a diagenetic redistribution of carbonate within an originally homogeneous sequence, without initial variations of stable oxygen isotope composition, then diagenesis would have resulted in an oxygen isotope profile which was the mirror-image of the ones which are actually observed. Therefore, a (completely) diagenetic origin of the carbonate–marl alternations is improbable.

If diagenesis has only emphasized pre-existing differences of carbonate content, then the original amplitude of the observed $\delta^{18}O$ fluctuations would have been diminished. Accordingly, the differences in $\delta^{18}O$ values between adjacent intervals then must have been larger originally. Considering the measured $\delta^{18}O$ values, and the rather small differences of carbonate content between adjacent carbonate and marly beds, diagenetic effects are not likely to have caused important changes.

4.1. Paleotemperatures

The rate in which ^{18}O and ^{16}O isotopes are incorporated within marine biogenic carbonate, relative to the oxygen isotope composition of sea-water, is dependent on temperature; with increasing temperature the $\delta^{18}O$ value of the carbonate decreases. During the Quaternary, the oxygen isotope composition of sea-water has been influenced by the storing of ice at the poles. Polar ice has $\delta^{18}O$ values as low as $-50\%_0$, and Shackleton and Kennett (1975) estimate that the formation of the Antarctic ice sheet has caused $\delta^{18}O$ of the ocean water to rise by about $1\%_0$.

Due to other causes, the oxygen isotope composition of sea-water has also fluctuated in times in which no polar ice caps existed. On the basis of the data of Claypool et al. (1980) and of Scholle & Arthur (1980), it is estimated that, during the Middle Cretaceous, the $\delta^{18}O$ value of ocean water was about $0.5-1\%_0$ less than it was immediately before the start of ice cap growth in the Tertiary. Assuming that $\delta^{18}O$ of ocean water was $1.5\%_0$ lower than at present, application of the equation of Shackleton & Kennett (1975) points to paleotemperatures for the analysed sections, varying from $19.3°C$ (for $\delta^{18}O = -2.4\%_0$) to $24.9°C$ (for $\delta^{18}O = -3.6\%_0$). In case of a $2\%_0$ lower value of the Middle Cretaceous ocean waters, the relative paleotemperatures would have been $17.1°C$ and $22.5°C$. Uncertainty about the exact oxygen isotope composition of the Middle Cretaceous oceans makes such estimates tentative. Anyhow, oxygen isotope data indicate fluctuations of temperature of a magnitude of 3 or 4 degrees, with *the carbonate-richer beds having been formed during periods with cooler water*. These differences of temperature do not apply to the very water surface, but to the entire part of the water column in which the bulk of the carbonate is formed, i.e. the photic zone.

5.   Stable carbon isotopes

5.1.   Carbonate in organic carbon-rich intervals

Irwin, Curtis & Coleman (1977) have put forward a model of burial diagenesis of organic matter in marine sediments. With the increase of burial depth they subsequently distinguish (1) bacterial oxidation, (2) sulphate reduction, (3) bacterial fermentation and (4) abiotic reactions. During (1), (2) and (4), carbon dioxide is set free with about the same carbon isotope composition as the organic matter from which it is derived (i.e. $\pm -25\%_0$). Only during bacterial fermentation, which starts after the consumption of sulphate, the CO_2 released is isotopically heavy, in the order of $+15 - +30\%_0$. CH_4 released at the same time may be as light as $-75\%_0$. Measurements of recent pore waters and young occurrences of natural gas support this idea; $\delta^{13}C$ CH_4 values as low as $-80\%_0$ are found, and $\delta^{13}C$ CO_2 as high as $+25\%_0$ (Faber, Schmitt & Stahl, 1978; Carothers & Kharaka, 1980; Schoell, 1980). Also the negative correlation of the sulphate ion content and $\delta^{13}C$ of HCO_3^- in oil-field pore waters, shown by data of Carothers & Kharaka (1980), indicates that only after the consumption of sulphate, heavy carbon dioxide is liberated. A 20% lowering of salinity and sulphate content of the ocean water as a result of evaporite deposition during the Early and Middle Cretaceous, suggested by Hay (1981), may have favoured an early start of bacterial fermentation, and the production of HCO_3^- with high $\delta^{13}C$ values during this period.

The heavy CO_2 released during bacterial fermentation, dissolves easily in the pore water. There, it may take part in dissolution–precipitation processes of sedimentary carbonate, and cause an increase of $\delta^{13}C$. On the other hand, the light CH_4-carbon produced during bacterial fermentation is *not* active in the anoxic environment where it is formed. It has to migrate to higher water levels, where it can be oxidized and can influence $\delta^{13}C$ values of the bulk carbon dioxide. This oxidation is a slow process; even under present conditions with well-ventilated deep waters, the bulk of the methane which escapes from the sea floor is oxidized at levels above a depth of 500 meters (Brooks, 1979).

The diagenetic influence of heavy carbon dioxide, produced during bacterial fermentation, upon the carbon isotope composition of carbonate was demonstrated by Campos & Hallam (1979). High $\delta^{13}C$ $CaCO_3$ values from sample station 1474 in the Black Sea (Deuser, 1972) are likely to be the result of the same process.

These data lead to the suggestion that *the high $\delta^{13}C$ values of carbonate in the organic carbon-rich intervals of Figure 1 do not reflect an original trend, but that heavy carbon dioxide, released during early diagenesis, has caused a diagenetic increase of $\delta^{13}C$ of the bulk carbonate.*

5.2. Organic matter

$\delta^{13}C$ values of organic matter in black shale intervals of Figure 1 are low, compared to those in adjacent carbonate-rich beds. Often, low $\delta^{13}C$ values are ascribed to the input of terrestrial organic matter. Although visible plant remains are scarce or absent, van Graas, Viets, de Leeuw & Schenck (1981) show that eolian input of terrestrial plant waxes may have contributed significantly to the organic matter in such pelagic sequences.

Our research indicates that carbonate-poor (black shale) intervals were deposited in times of low surface productivity. The supply of terrestrial organic matter via the air is not dependent on the productivity of the surface waters in which it ends up. Thus *the relative proportion of terrestrially derived input in the total amount of organic matter is likely to be larger in parts of the sequence that were formed in periods of low production, than in other parts which were formed in times of high production. This relation might explain the observed variations of $\delta^{13}C$ of the organic matter.*

$\delta^{13}C$ of organic matter seems to be stable after fossilization. Pyrolysis of organic matter at 500°C for 11 days resulted in only small differences in $\delta^{13}C$ between the residue and the original organic matter, the maximum difference being 0.9‰ (Chung & Sackett, 1979). All evidence is compatible with the assumption that $\delta^{13}C$ values of bulk samples of organic matter do reflect the original values.

If the relatively large variations of $\delta^{13}C$ of organic matter indeed reflect a varying rate of contribution of terrestrial organic matter, it is not likely that the effects of other processes, discussed below with respect to $\delta^{13}C$ of the carbonate, would be noticeable.

5.3. Carbonate in intervals with no or little organic matter

In contrast to the dark-coloured marls, $\delta^{13}C$ values of carbonate in light-coloured marly intervals are not more positive than in adjacent carbonate-rich beds, but, on the contrary, they show a tendency to deviate negatively from the general trend (Figures 2, 4). They are assumed not to have changed as much as those within the black shales, and their fluctuations are assumed to reflect original variations of the environment in which they have been formed, i.e. the photic zone. The differences measured, are not very significant. However, it should be realized that fluctuations in the stable isotope composition of inorganic carbon in the open ocean, where the carbonate was formed, are also likely to have been small. In this context, the stable carbon isotope composition of carbonate in the studied sequences will be discussed below.

6. Environment of deposition

The carbonate fraction of the studied sediments consists mainly of calcareous nannoplankton and foraminifera. As shown above, oxygen isotope data indicate that the carbonate of the carbonate-richer beds was formed in relatively cool water, at maximum 3–4° colder than the water in which the carbonate of the marly intervals was formed.

The sediments were deposited at a depth of 1–2 km (Wonders, 1980), i.e. above the carbonate compensation depth, which is supposed to have been at 3–4 km during the Middle Cretaceous (van Andel, 1975). Therefore, the fluctuations of carbonate content may well reflect variations of surface productivity.

Although the organic carbon-richness of anoxic sediments suggests highly productive surface waters, this relation is not necessary. Assuming that about 10% of the primary produced organic matter reaches the sediment surface, a primary productivity at the surface of less than 25 g m^{-2} yr^{-1} of organic carbon is sufficient to produce the amount of organic matter in black shales such as found for example in the Bonarelli Level in the Apennines, which is well-known for its high content of organic matter (up to 15%; estimated rate of deposition about 0.2 cm per 1000 years). This figure may be compared to the primary productivity in low productive parts of the recent oceans (50–100 g m^{-2} yr^{-1}, Bolin, Degens et al., 1979). Thus, deposition of black shales is not necessarily the result of a great organic productivity and of an high input of organic matter, but a lack of supply of oxygen to deep water is at least as important. In the present-day Black Sea, oxygen-deficient sea-floor regions are also characterized by low organic productivity at the surface (Shimkus & Trimonis, 1974).

On the basis of organic geochemical results, van Graas et al. (1981) also conclude that productivity in the water column, which supplied the sediments shown in Figure 4, has indeed been low.

It follows that, in the case of the sediments discussed here, the carbonate content is the best measure of productivity in surface waters during the time of deposition.

Supply of nutrients, which is the prerequisite for organic activity, was therefore likely to be relatively great in times during which the carbonate-rich intervals were formed. Oxygen isotope data point to a relatively low temperature of formation of the carbonate of these intervals. Therefore, upwelling of cool and nutrient-rich water from deeper levels is the most logical explanation for an increased nutrient supply. Due to upwelling, the thermocline may be as shallow as 50 m below the water surface near the equator, whereas it is situated at deeper levels (100–200 m and more) at higher latitudes (Defant, 1961). In the case of the Moria sediments, the rate of upwelling and productivity must have fluctuated regularly above the place of deposition.

6.1. Cyclicity

It has been shown that the carbonate–marl alternations within the Scaglia Bianca Fm. have a periodicity of about 20 000 years (de Boer & Wonders, 1981). The value of 20 000 years, found in Cretaceous pelagic sediments in the Apennines as well as in other stratigraphic intervals at other places, suggests a relationship with the influence of astronomical parameters upon climate. Indeed, the pattern of the variation of thickness of the successive carbonate beds corresponds closely to the pattern of variation of eccentricity and precession effects (de Boer & Wonders, 1981; de Boer, 1982).

Fossil sediments with rhythms of the order of 20 000 years have often been deposited at paleolatitudes of about 30°N or S (Schwarzacher, 1954; van Houten, 1962; Fischer, 1964; Arthur & Fischer, 1977; McCave, 1979; etc.). Berger (1978) shows that the astronomical parameters precession and eccentricity cause a continuous northward and southward shifting of the caloric equator over the geographic equator with an average periodicity of about 21 000 years. The caloric equator is the line of maximum (solar) insolation, not necessarily coinciding with the geographic equator. Another phenomenon at this latitude is the tropical upwelling zone, which is reflected in the shallow depth of the thermocline (cf. Defant, 1961). Its asymmetry with respect to the geographic equator, is obviously related to the position of the caloric equator.

The regular shifting of the caloric equator from (at maximum) 10°N to 10°S (Berger, 1978) may cause areas around paleolatitudes of 20–35° to fall alternately inside and outside the influence of the equatorial zone of upwelling and high productivity. Rhythmic variations of circulation intensity, and of supply of nutrients to the surface waters may then cause a periodic increase and decrease of productivity of carbonate and of oxygen renewal in deep water, leading to the formation of sequences of alternating carbonate-rich, and marly/organic carbon-rich deposits.

Astronomical parameters seem to trigger rather than to control the growth and melting of polar ice caps; therefore, fluctuations of the global climate during the Quaternary have been relatively large as compared to periods in which no polar ice caps existed (cf. Hays, Imbrie & Shackleton, 1976; Oerlemans, 1980). No indications exist that ice caps were present during the Cretaceous. Therefore, short term (in the order of ten thousands of years) variations of climate have probably been much smaller than during the subrecent history (de Boer, 1982).

7. Discussion: factors influencing $\delta^{13}C$ in the photic zone

7.1. Black shale formation and productivity rates

The $^{13}C/^{12}C$ ratio of marine carbon increases, when a prolonged (in the order of millions of years) net distraction of organic matter from the active carbon cycle occurs; such an increase during the Middle Cretaceous is reflected by, e.g., fossil carbonates (Scholle & Arthur, 1980; Veizer, Holser & Wilgers, 1980).

However, in the short term, removal of organic matter from the photic zone (where the bulk of the biogenic sediment is synthesized) into pelagic sediments can have only a very small impact upon the mean isotope ratio of the oceanic carbon reservoir: sedimentation rates of pelagic sediments such as those of the Mid-Cretaceous series in the Apennines are in the order of 1 cm per 10^3 yrs. In the case of black shales with a 5% content of organic carbon, about 0.5 g m^{-2} of carbon would have been fossilized annually as organic matter; a very small quantity if one considers that this is less than a permille of the amount of carbon present within the photic zone in a column with basic area of 1 m^2. Moreover, the residence time of carbon in the oceanic surface layer is estimated to be in the order of only 5–10 years (Kester & Pytkovitch, 1977), and the exchange of carbon dioxide with the atmosphere is of the order of 100 gr C per m^2 per year. Therefore, in the short term, replacement of carbon in surface waters must have been sufficiently rapid to compensate for an increase of $\delta^{13}C$ caused by the distraction of organic matter due to black shale formation.

This implies that, in contrast to long term effects, the storage of organic matter in pelagic settings on short term can hardly have influenced $\delta^{13}C$ of inorganic carbon in the oceanic surface layer. On the contrary, it is suggested that an increase of $\delta^{13}C$ in surface waters, because of the removal of organic carbon from the photic zone, should not be expected to occur in times of black shale deposition, but in times with good water circulation, during which relatively highly productive surface waters supply abundant organic matter for transportation to deep water.

Whether oxidation of the sinking organic matter occurs or not, it is not, or only indirectly, related to the surface production, since it depends on the oxidation state of deep water. At present the circulation velocity of ocean water is measured in hundreds of years. C 14 ages measured on organic matter suspended in deep water, as reported by various authors, vary between 1000 and 3000 years. Circulation velocity during the Cretaceous must have been slower than at present. Therefore, the light carbon of organic matter which was liberated by oxidation in deep water, was brought back to the surface waters slowly, with a delay of thousands of years. Irrespective of the oxygenation state of bottom waters, the transport of organic matter from the surface layer into the deep must have had its greatest influence upon $\delta^{13}C$ in surface waters during times of high productivity, i.e. during the formation of the carbonate-richer beds.

7.2.   Upwelling and productivity

The correlation between temperature and $\delta^{13}C$ of biogenic products observed in recent oceanic environments is probably a derivate from the combination of the facts that (1) high productivity is often caused by the upwelling of cold, deep, nutrient-rich water, which (2) is generally depleted in ^{13}C, and (3) that exhaustion of the HCO_3^- stock during high productivity, and consumption of light dissolved gaseous CO_2 ($\delta^{13}C = \pm -10\%_0$) instead of HCO_3^-, may occur in case of a high surface productivity. However, if HCO_3^- is sufficiently available, this latter effect is absent (Goodney et al., 1980). Primary productivity in the sluggishly circulating Cretaceous oceans is assumed to have been so low that the concentration of HCO_3^- has been continuously sufficient for photosynthetic processes, and that consumption of dissolved CO_2 has been absent or negligible in the surface waters under consideration here.

At present, $\delta^{13}C$ of inorganic carbon in deep water and in surface waters may differ more than a permille, the values in deep water being the lower ones, because of the decay of organic matter. During the Cretaceous, this difference has been less, of the order of $0.5\%_0$ (Berger, 1979), indicative of the low oceanic productivity during this period. Therefore, the negative influence of upwelling upon $\delta^{13}C$ of the surface waters must have been less than at present. In addition, the low circulation velocity of ocean water as compared to the present situation, must also have caused the influence of upwelling upon $\delta^{13}C$ in the surface waters to have been less than at present. The lowering of the mean $\delta^{13}C$ value of inorganic carbon in the surface waters because of upwelling of ^{13}C depleted waters is counteracted by the production and transport of organic matter to the deep, especially in times of high productivity.

Due to the ^{13}C depletion of organic matter ($\delta^{13}C \cong -25\%_0$), increased production of organic carbon and transport to deep water may lead to an enrichment of ^{13}C in the upper part of the water column, particularly during periods of high productivity. This latter effect is of instantaneous influence whereas the isotope ratio of upwelling deep water has an intermediate value as a result of the combination of the effects of high and low productive periods. Therefore, it may be expected that the effect of sinking of ^{13}C-depleted organic matter during highly productive periods had a greater influence upon $\delta^{13}C$ of the surface waters than had the upwelling of ^{13}C depleted deep water.

If varying rates of upwelling and productivity were the only factors involved, then the long residence time and mixing of deep waters would result in mediocre $\delta^{13}C$ values of upwelling deep water. Superimposed would be a more pulsating influence of varying rates of distraction of light organic carbon by varying rates of productivity, leading to a rise of $\delta^{13}C$ in highly productive periods and to a lowering in times of low productivity.

7.3. Fractionation during the formation of carbonate

The isotopic effect between carbonate and dissolved bicarbonate is temperature dependent, $+0.04‰$ per $°C$ (Emrich, Ehhalt & Vogel, 1970). In the case of temperature fluctuations of 3–4°, as suggested by the above oxygen isotope data, the effect would be about $0.1‰$, the marly beds having the higher values. As foraminifera form only a minor portion of the carbonate, species dependent deviations of $\delta^{13}C$ from isotopic equilibrium within forams (Shackleton, Wiseman & Buckley, 1973) are not considered to be important for the values measured on the bulk samples. Deviating carbon isotope ratios in calcareous nannofossils—which make up the bulk of the analysed samples—are, at present, especially dependent on high rates of primary productivity which was shown above to have been continuously low during deposition of the discussed sediments; so, this effect seems negligible. Thus for the sediments under discussion, only the above small temperature effects of $+0.04‰$ per $°C$ seems to be important.

7.4. Fractionation over the air–water interface

In equilibrium conditions, $\delta^{13}C$ of atmospheric CO_2 is about 7–10‰ less than that of HCO_3^- dissolved in sea water (Emrich, Ehhalt & Vogel, 1970; Mook, Bommerson & Staverman, 1974). This difference is dependent on temperature: for each degree colder, it increases by about $0.1‰$ (at normal temperatures). A lowering of the temperature thus leads to a net transfer of ^{13}C from the atmosphere into the surface layer of the ocean, and to an increase of $\delta^{13}C$ in surface waters.

It was concluded above that a relatively low mean temperature of sea water in periods with upwelling, caused the high $\delta^{18}O$ values of carbonate in carbonate-richer intervals.

Thus it would be attractive to relate the relatively positive values of $\delta^{13}C$ of carbonate in the carbonate-rich intervals to a lowering of the surface water temperature and to a related net transfer of ^{13}C from the atmosphere into the ocean. However, the upwelling of cool deep water is linked to the position of the caloric equator. Both factors, the upwelling of cool water, and the insolation maximum at the caloric equator, have an opposite effect upon the mean temperature at the water surface. Inspection of various temperature distribution maps gives the idea that upwelling does not cause a significant lowering of the temperature at the very water surface in the present-day tropics. It therefore is improbable that it yet could have done so in the slowly circulating Middle Cretaceous oceans. Thus, in times of upwelling, in which the photic zone as a whole did have a relatively low mean temperature at the paleolatitude of the studied sediments (about 30°N), the temperature at the very surface of the water probably has been slightly higher than in other periods in which upwelling occurred at more southerly latitudes. This may have led to a relative decrease, instead of a relative increase, of $\delta^{13}C$ of inorganic carbon in surface waters in times in which the caloric equator and the tropical upwelling zone did have a northerly position.

The temperature dependence of carbon isotope fractionation between water and air, however, may offer an explanation for the strong negative shift of $\delta^{13}C$ over the

Cretaceous–Tertiary boundary, observed in many places. The event seems to have covered a very short time interval and to have resulted in a global warming (see Christensen & Birkelund, 1979). Such a warming may well have caused a transfer of ^{13}C from the surface layer of the ocean into the atmospheric realm. In combination with other processes, such as a reduction of the terrestrial biosphere, this may have lead to the observed ^{13}C depletion of marine carbonates of that age.

7.5. Net effect

Of the above discussed processes, only varying rates of removal of organic matter to deep water, in proportion to the rate of primary productivity, would explain a sequence of data such as shown in Figure 2. The other processes, which have been discussed above, would have an opposite effect. It is therefore suggested that during periods with a relatively high productivity, the distraction of light organic carbon from the surface waters led to a small but noticeable increase of $\delta^{13}C$.

The above model, suggesting shifts of the tropical upwelling and high productivity zone to be the cause of the rhythmic carbonate-marl alternations, should be considered to be tentative. Probably the situation was more complex. One can propose other models, in which temperature dependent fractionation of carbon isotopes over the air-water interface indeed would add to the effect seen in Figure 2.

8. Conclusions

The low $\delta^{13}C$ values of organic matter in the black shale intervals discussed here, seem to be the result of a, not absolutely, but relatively high input of, probably wind-blown, terrestrial organic matter during times of low organic production in the water column.

High $\delta^{13}C$ values of carbonate in black shale intervals are due to the precipitation of heavy carbon isotopes from heavy HCO_3^-, resulting from bacterial fermentation.

The rhythmicity observed in the studied sequences can be ascribed to a regular north- and southward shifting of the caloric equator and of oceanic current patterns. This shifting is the result of the influence of astronomical parameters and the periodicity is about 20 000 years. On latitudes of about $30°$, this results in a regular increase and decrease of upwelling, nutrient supply, replenishment of oxygen in deep water, and water temperature. During periods in which the place of deposition was in the sphere of influence of upwelling and high productivity, the mean temperature within the part of the water column, in which biogenic carbonate was formed, was relatively low, productivity was high, and bottom waters were well-oxygenated.

In times in which the caloric equator was situated in the other hemisphere, primary production was low. In such periods, oxygen replenishment in deep water was sometimes reduced to the extent that anoxic sediments (black shales) could be formed. The carbonate content of the sediment thus seems to be the better tool for the reconstruction of variations of productivity during deposition of sequences such as described.

Stable carbon isotope values of carbonate in parts of the studied sequences, that were formed under continuously oxygenated conditions in deep water, tend to be slightly higher in the carbonate-rich than in the marly beds. The only reasonable explanation found is a relation between primary productivity and the amount of ^{13}C-depleted organic matter which is removed by sinking into deep water.

Acknowledgments

I thank Drs G. van Graas, P. Marks, and A. A. H. Wonders for discussions and cooperation in the field. The help of Mr. J. A. N. Meesterburrie in the analyses of stable isotope ratios is gratefully acknowledged. Drs J. A. McKenzie, E. Nickel, W. Salomons, and G. J. van der Zwaan are thanked for reading and commenting on an earlier version. The manuscript has benefited greatly from the critical comments of Professor Dr. P. Marks.

References

Arthur, M. A. 1977. Sedimentology of the Gubbio sequence and its bearing on palomagnetism. *Mem. Soc. Geol. Italia* **15**, 9–20.

Arthur, M. A. 1979. North Atlantic black shales: the record at site 398 and a brief comparison with other occurrences. *Initial Reports of the Deep Sea Drilling Project* **47** (2), 719–738.

Arthur, M. A. & Fischer, A. G. 1977. Upper Cretaceous—Paleocene magnetic stratigraphy at Gubbio, Italy. I. Lithostratigraphy and sedimentology. *Geological Society of America Bulletin* **88**, 367–371.

Berger, A. L. 1978. Long-term variations of caloric insolation resulting from the earth's orbital elements. *Quaternary Research* **9**, 139–167.

Berger, W. H. 1979. Impact of deep-sea drilling on paleoceanography. In *Deep Sea Drilling Results in the Atlantic Ocean:* continental margins and paleoenvironment (Eds M. Talwani, W. Hay & W. B. F. Ryan), Maurice Ewing Series 3, American Geophysical Union, 297–314.

Bolin, B., Degens E. T., Duvigneaud, P. & Kempe, S. 1979. The global biogeochemical carbon cycle. In *The global carbon cycle* (Eds B. Bolin, E. T. Degen, S. Kempe & P. Ketner). Wiley: Chichester, 1–53.

Brooks, J. M. 1979. Deep methane maxima in the Northwest Caribbean Sea: possible seepage along Jamaica Ridge. *Science* **206**, 1069–1071.

Campos, H. S. & Hallam, A. 1979. Diagenesis of English Lower Jurassic limestones as inferred from oxygen and carbon isotope analysis. *Earth and Planetary Science Letters* **45**, 23–31.

Carothers, W. W. & Kharaka, Y. K. 1980. Stable carbon isotopes of HCO_3^- in oil-field waters—implication for the origin of CO_2. *Geochimica et Cosmochimica Acta* **44**, 323–332.

Christensen, W. K. & BirkelunC, T. (Eds) 1979. *Cretaceous–Tertiary Boundary Events; Symposium*. Vol. 2 proceedings.

Chung, H. M. & Sackett, W. M. 1979. Use of stable carbon isotope compositions of pyrolytically derived methane as maturity indices for carbonaceous materials. *Geochimica et Cosmochimica Acta* **43**, 1979–1988.

Claypool, G. E., Holser, W. T., Kaplan, I. R., Sakai, H. & Zak, I. 1980. The age curves of sulfur and oxygen isotopes in marine sulfate and their mutual interpretation. *Chemical Geology* **28**, 199–260.

de Boer, P. L. 1982. Cyclicity and the storage of organic matter in Middle Cretaceous pelagic sediments. *Cyclic and Event Stratification* (Eds G. Einsele & A. Seilacher). Springer: New York.

de Boer, P. L. & Wonders, A. A. H. 1981. Milankovitch parameters and bedding rhythms in Umbrian Middle Cretaceous pelagic sediments. *International Association of Sedimentologists (I.A.S.) 2nd Eur. Meeting, Bologna Abstracts*, 10–13.

Defant, A. 1961. Physical Oceanography. Vol. 1. Oxford: Pergamon, 729pp.

Deuser, W. G. 1972. Late-Pleistocene and Holocene history of the Black Sea as indicated by stable-isotope studies. *Journal of Geophysical Research* **6**, 1071–1077.

Emrich, K., Ehhalt, D. H. & Vogel, J. C. 1970. Carbon isotope fractionation during the presipitation of calcium carbonate. *Earth and Planetary Science Letters* **8**, 363–371.

Faber, E., Schmitt, M. & Stahl, W. 1978. Carbon isotope analyses of head space methane from samples of leg 42b, sites 379, 380 and 381. *Initial Reports of the Deep Sea Drilling Project* **42b**, 667–672.

Fischer, A. G. 1964. The Lofer Cyclothems of the Alpine Triassic. *Kansas Geological Survey Bulletin* **169**, 107–149.

Fischer, A. G. & Arthur, M. A. 1977. Secular variations in the pelagic realm. *Society of Economic Paleontologists and Mineralogists; Special Publication* (Eds H. E. Cook & P. Enos) **25**, 19–50.

Frakes, L. A. 1979. *Climates throughout Geologic Time*. Elsevier, 310pp.

Goodney, D. E., Margolis, S. V., Dudley, W. C., Kroopnick, P. & Williams, D. F. 1980. Oxygen and carbon isotopes of recent calcareous nannofossils as paleoceanographic indicators. *Marine Micropaleontology* **5**, 31–42.

Hay, W. W. 1981. Sedimentological and geochemical trends resulting from the breakup of Pangea. Acta 1981. *Proceedings 26th International Geological Congress; Geology of Oceans Symposium, Paris, July 1980*, 135–147.

Hays, J. D. & Pitman, W. C. 1973. Lithospheric plate motion, sea level changes and ecological consequences. *Nature* **246**, 18–22.

Hays, J. D., Imbrie, J. & Shackleton, N. J. 1976. Variations in the Earth's orbit: pacemaker of the ice ages. *Science* **194**, 1121–1132.

Irwin, H., Curtis, C. & Coleman, M. 1977. Isotopic evidence for source of diagenetic carbonates formed during burial of organic-rich sediments. *Nature* **269**, 209–213.

Jenkyns, H. C. 1980. Cretaceous anoxic events: from continents to oceans. *Journal of the Geological Society London* **137**, 171–188.

Kester, D. R. & Pytkovitch, R. M. 1977. Natural and anthropogenic changes in the global carbon dioxide system. In *Global Chemical Cycles and Their Alterations by Man* (Ed. W. Stumm), Report Dahlem Workshop: Berlin 1976, 99–120.

Kroopnick, P. M., Margolis, S. V. & Wong, C. S. 1977. $\delta^{13}C$ variations in marine carbonate sediments as indicators of the CO_2 balance between the atmosphere and oceans. *The fate of fossil fuel CO_2 in the oceans* (Eds N. R. Andersen & A. Malahoff) 295–320.

McCave, I. N. 1979. Depositional features of organic-carbon-rich black and green mudstones at D.S.D.P. Sites 386 and 387, Western North Atlantic. *Initial Reports of the Deep Sea Drilling Project* **43**, 411–416.

McKenzie, J. A., Bernouilli, D. & Garrison, R. E. 1978. Lithification of pelagic-hemipelagic sediments at DSDP site 372: oxygen isotope alteration with diagenesis. *Initial Reports of the Deep Sea Drilling Project* **42A**, 473–478.

Mook, W. G., Bommerson, J. C. & Staverman, W. H. 1974. Carbon isotope fractionation between dissolved bicarbonate and gaseous carbon dioxide. *Earth and Planetary Science Letters* **22**, 169–176.

Oerlemans, J. 1980. Model experiments on the 100,000 year glacial cycle. *Nature* **287**, 430–432.

Ryan, W. B. F. & Cita, M. B. 1977. Ignorance concerning episodes of ocean-wide stagnation. *Marine Geology* **23**, 197–215.

Schlanger, S. O. & Jenkyns, H. C. 1976. Cretaceous oceanic anoxic events: causes and consequences. *Geologie en Mijnbouw* **55**, 179–184.

Schoell, M. 1980. The hydrogen and carbon isotopic composition of methane from natural gases of various origins. *Geochimica et Cosmochimica Acta* **44**, 649–661.

Scholle, P. A. & Arthur, M. A. 1980. Carbon isotope fluctuations in Cretaceous pelagic limestones: potential stratigraphic and petroleum exploration tool. *American Association of Petroleum Geologists Bulletin* **64**, 67–87.

Schwarzacher, W. 1954. Die Grossrhythmik des Dachsteinkalkes von Lofer. *Tchermaks Mineralogische und Petrographische Mitteilungen* **4**, 44–54.

Shackleton, N. J. & Kennett, J. P. 1975. Paleotemperature history of the Cenozoic and the initiation of Antarctic glaciation: oxygen and carbon isotope analyses in DSDP sites 277, 279, 281. *Initial Reports of the Deep Sea Drilling Project* 743–755.

Shackleton, N. J., Wiseman, J. D. H. & Buckley, H. A. 1973. Non-equilibrium isotopic fractionation between seawater and planktonic foraminiferal test. *Nature* **242**, 177–179.

Shimkus, K. M. & Trimonis, E. S. 1974. Modern sedimentation in Black Sea. (Eds E. T. Degens & D. A. Ross). The Black Sea—Geology, chemistry and biology. *American Association of Petroleum Geologists Memoirs* **20**, 249–278.

Thiede, J. & van Andel, T. H. 1977. The paleonvironment of anaerobic sediments in the late Mesozoic South Atlantic Ocean. *Earth and Planetary Science Letters* **33**, 301–309.

van Andel, T. H. 1975. Mesozoic/Cenozoic calcite compensation depth and the global distribution of calcareous sediments. *Earth and Planetary Science Letters* **26**, 187–194.

van Graas, G., Viets, T. C., de Leeuw, J. W. & Schenck, P. A. 1981. Origin of the organic matter in a Cretaceous "Black Shale" deposit in the central Apennines (Italy). *Advances in Geochemistry* (Ed. M. Bjoroy).

van Houten, F. B. 1962. Cyclic sedimentation and the origin of analcim-rich upper Triassic Lockatong, west-central New Jersey and adjacent Pennsylvania. *American Journal of Science* **260**, 561–576.

Veizer, J., Holser, W. T. & Wilgus, C. K. 1980. Correlation of $^{13}C/^{12}C$ and $^{34}S/^{32}S$ secular variations. *Geochimica et Cosmochimica Acta* **44**, 579–587.

Weissert, H., McKenzie, J. & Hochuli, P. 1979. Cyclic anoxic events in the early Cretaceous Tethys Ocean. *Geology* **7**, 147–151.

Wonders, A. A. H. 1980. Middle and Late Cretaceous planktonic foraminifera of the western Mediterranean Area. *Utrecht Micropaleontological Bulletins* **24**, 157 pp.

8. Organic Geochemical Indicators for Sources of Organic Matter and Paleoenvironmental Conditions in Cretaceous Oceans

B. R. T. Simoneit*

Institute of Geophysics and Planetary Physics, University of California

D. H. Stuermer

Lawrence Livermore National Laboratory, University of California

Cretaceous black shales have an extensive distribution on various continental areas, in the Atlantic and other isolated areas of the world oceans. They are organic-rich sediments which were deposited in anoxic environments (possibly worldwide oceans) under various sedimentary conditions. The literature is briefly reviewed and some type areas have been examined using organic geochemical indicators to assess the sources of the organic matter and the paleoenvironmental conditions of sedimentation. These consisted of the eastern Angola Basin, the Cape Verde Basin and eastern North Atlantic, the western North and South Atlantic, and the Pacific Oceans.

The organic geochemical indicators that were applied consisted of lipid and kerogen analyses. The lipids were characterized by their stable carbon isotope content, homolog and hump distributions and molecular markers, thus delineating the terrigenous versus autochthonous sources and the paleoenvironmental preservation of the lipids. The kerogens were characterized by their elemental composition, stable carbon isotope, electron spin resonance and vitrinite reflectance data. These parameters confirm that the organic matter in sediments of the Cretaceous oceans was derived from several different sources and was preserved under various paleoenvironmental conditions.

1. Introduction

Recent results of the Deep Sea Drilling Project reveal that the occurrence of organic carbon-rich marine sediments of mid-Cretaceous age is much more widespread than previously realized (Arthur & Schlanger, 1979; Ryan & Cita, 1977; Thiede & van Andel, 1977; Fischer *et al.*, 1977; Schlanger & Jenkyns, 1976). Evidence further suggests that these deposits are the result of the preservation of a variety of organic source materials by various mechanisms depending on location in the World Ocean (Tissot *et al.*, 1979, 1980; Arthur & Schlanger, 1977; Ryan & Cita, 1977). Apparently, two mechanisms have been identified that caused the anoxic conditions responsible for the preservation of the organic carbon in the sediments. These are (1) stratified water with basin-wide anoxic conditions in the bottom waters at Atlantic and Indian Ocean sites, and (2) an oxygen minimum or depleted zone in the water column intersecting topographic highs in the Pacific and Atlantic Oceans (Schlanger & Jenkyns, 1976; Thiede & van Andel, 1977; Ryan & Cita, 1977; Demaison & Moore, 1980). Basin-wide anoxic conditions are not unusual occurrences in the geologic past (Degens & Stoffers, 1976) and, indeed, several modern-day examples exist in the Black Sea (Degens & Ross, 1974), the Mediterranean Sea, the Cariaco Trench (Ryan & Cita, 1977; Degens & Stoffers, 1976) and in the basins of southern California

---

* Present address: School of Oceanography, Oregon State University.

(Emery, 1960). Examples of the intersection of the oxygen-minimum layer with continental shelf sediments is observed now off the Peru–Chile coast and off the India–Pakistan coast (Arthur & Schlanger, 1979). However, the conditions of climate, topography, oceanic circulation, and productivity that contributed to such widespread, worldwide occurrences of these organic carbon-rich deposits remains an intriguing problem. Solutions to this problem have practical implications for petroleum exploration (Tissot *et al.*, 1979, 1980; Arthur & Schlanger, 1979) and for understanding the effects of the present-day increase of atmospheric CO_2 concentrations caused by the combustion of fossil fuels (Ryan & Cita, 1977).

Determining the source of the organic matter in these mid-Cretaceous organic carbon-rich sediments also has important consequences for understanding paleoenvironmental conditions (Gartner, 1979; Thiede & van Andel, 1977) and determining their petroleum source potential (Philippi, 1974; Tissot *et al.*, 1974; Dow, 1977). Studies to date suggest an extremely variable source for these organic deposits (Tissot *et al.*, 1979, 1980; Bogolyubova & Timofeev, 1978). Our results using geochemical indicators give new evidence for several varied sources in depositional environments including the eastern Angola Basin, the Cape Verde Basin, the eastern North Atlantic, the western North and South Atlantic, and the mid-Pacific.

2. Experimental

The basic experimental procedures for the sample preparation, extraction, fractionation and derivatization prior to analysis were detailed elsewhere (e.g. Simoneit, 1975, 1980*c*; Simoneit *et al.*, 1972, 1973). The lipids were separated into hydrocarbon, fatty acid (as methyl esters) and ketone fractions, which were analyzed by gas chromatography (GC) and gas chromatography–mass spectrometry (GC/MS). The GC operating conditions and GC/MS operating parameters were as described elsewhere (Simoneit, 1975, 1980*c*).

The humic substances were separated by the procedure of Stuermer *et al.* (1978) and the residual kerogen was concentrated by repeated treatments with HCl and HF. The stable carbon isotope analyses were carried out by the method described (Kaplan *et al.*, 1970; Stuermer *et al.*, 1978), using Chicago PDB as the reference standard and the results are expressed in the δ notation. The procedures for the determinations of the electron spin resonance (ESR) spectra and the vitrinite reflectance data have been described (Simoneit *et al.*, 1981; Peters *et al.*, 1979).

3. Results and discussion

The sample descriptions and analytical results for the lipids are found in Table 1 and for the kerogens the data are given in Table 2. The sample locations are shown in Figure 1.

3.1. Lipids

Lipids of marine sediments consist of homologous compounds attributable to autochthonous sources, with varying amounts of allochthonous influx contributed from continental sources (Simoneit, 1975, 1978*b*, 1980*a*). This is also the case for the organic-rich shales of the Cretaceous.

The samples from the north-eastern Atlantic (Sites 138, 367, 368, 397, 398 and 402) are of mid-Cretaceous ages and the lipids are of a predominantly terrigenous origin. An example of the *n*-alkane and *n*-fatty acid distributions for a sample from site 367 is

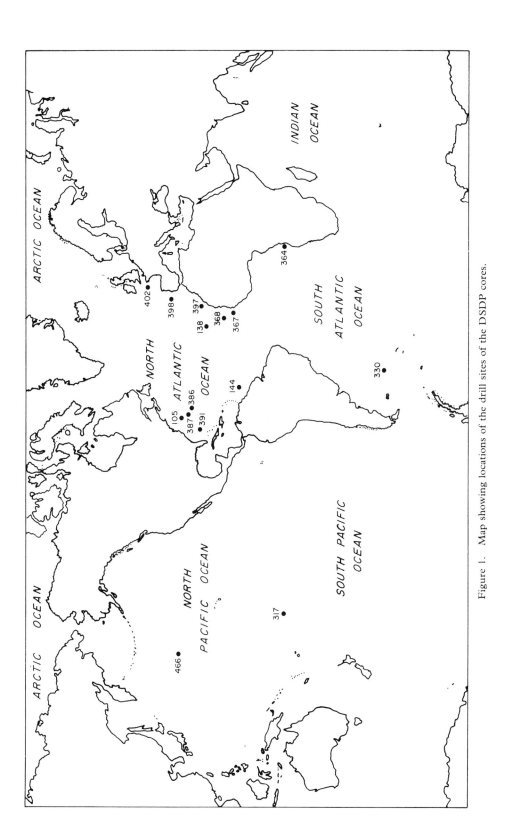

Figure 1. Map showing locations of the drill sites of the DSDP cores.

Table 1. Sample locations and analytical results for carbon and lipid content

Sample Number	Designation*	Location	Depth below seabed (m)	Water depth (m)	Geologic age	Carbon† Total (%)	Organic (%)	CaCO₃ (%)†	n-alkanes (µg/g)	CPI§	Max.||	Pr/Ph¶	Lipids n-fatty acids (µg/g)	CPI§	Max.||	δ¹³C, total fraction (‰)‡	Reference**
1	11-105-11-2 (10–138 cm)	34 54′ N 69 10′ W	310.0	5251	L. Cret.	5.7	3.6	18	n.d.	1.2	17,29	0.7	n.d.	3.5	16,24	n.d.	a
2	14-138-6-3 (49–50 cm)	25 55.4′ N 25 33.8′ W	428.5	5288	M. Cret.	16.8	16.8	0	3000	1.07	17,27	n.d.	1400	1.2	16,24	n.d.	b
3	14-144-4-2 (20–21 cm)	09 27.2′ N 54 20.5′ W	214.7	2957	U. Cret.	13.1	10.3	24	n.d.	—	—	—	n.d.	—	—	n.d.	b
4	14-144-4-3 (14–15 cm)	09 27.2′ N 54 20.5′ W	216.1	2957	U. Cret.	15.1	6.5	71	2200	1.8	17,29	n.d.	600	1.4	16,26	−27.5	b
5	14-144A-5-1 (114–116 cm)	09 27.2′ N 54 20.5′ W	181.0	2957	U. Cret.	16.8	11.1	48	200	1.7	17,27	n.d.	180	1.5	16,24	−27.2	b
6	14-144A-6-1 (100–101 cm)	09 27.2′ N 54 20.5′ W	190.1	2957	U. Cret.	17.0	10.4	55	950	1.7	17,23	1.04	1700	1.4	16,26	n.d.	b
7	33-317A-16-2 (131–134 cm)	11 00.9′ S 162 15.8′ W	680.3	2598	Cret. (Apt.)	25.3	24.9	3	1400	1.8	18,22,31	0.8	1300	7.2	16,24,28	−23.3	h
8	36-330-4-2 (120–126 cm)	50 55.2′ S 46 53.0′ W	274.2	2626	Cret. (Neocom.)	3.6	3.3	3	3.5	1.9	19,23,31	0.5	4	7.3	16,28	n.d.	i
9	36-330-10-1 (98–102 cm)	50 55.2′ S 46 53.0′ W	405.5	2626	U. Jur.	3.1	3.1	0	25	1.6	17,19,29	0.8	8	3.1	16,24	n.d.	i
10	36-330-14-4 (127–136 cm)	50 55.2′ S 46 53.0′ W	524.3	2626	M. Jur.	1.8	1.8	0	n.d.	—	—	—	n.d.	—	—	−24.1	i
11	40-364-24-1 (0–10 cm)	11 34.4′ S 11 58.3′ E	672.5	2448	Cret. (Tur.)	11.5	10.3	10	5.2	1.7	23,29	0.5	5.3	2.3	16,24	−24.2	e
12	40-364-43-1 (130–133 cm)	11 34.4′ S 11 58.3′ E	1034.8	2448	Cret. (Apt.)	12.2	7.8	36	220	1.2	17,22,29	0.6	0.5	4.4	16,22	−26.5	e
13	41-367-19-4 (10–15 cm)	12 29.2′ N 20 02.8′ W	649.1	4748	Cret. (Cenom.)	11.3	9.0	18	1.7	2.9	17,31	0.3	0.4	5.3	16,24	−27.5	d
14	41-368-58-2 (106–114 cm)	17 30.4′ N 21 21.2′ W	923.6	3367	Cret. (Tur./Alb.)	1.0	0.7	3	220	1.14	23,31	0.52	50	5.2	16,22	−27.0	c
15	41-368-63-1 (48–53 cm)	17 30.4′ N 21 21.2′ W	978.5	3367	Cret. (Tur./Alb.)	4.7	4.2	4	5300	1.04	17,19,29	1.02	1100	2.9	16,22	−28.2	c

	DSDP sample[*]	Lat./Long.			Age											δ¹³C[‡]	Ref.[**]
16	41-368-63-2 (127–132 cm)	17 30.4′ N 21 21.2′ W	980.8	3367	Cret. (Tur./Alb.)	6.6	6.6	1	1540	1.21	15,19,29	0.80	850	1.25	14,15,18,24	−28.5	c
17	41-368-63-3 (110–114 cm)	17 30.4′ N 21 21.2′ W	982.1	3367	Cret. (Tur./Alb.)	8.6	7.1	12	3250	1.11	17,27	0.55	1140	2.3	16,26	n.d.	c
18	43-386-43-3 (138–141 cm)	31 11.2′ N 64 14.9′ W	740.7	4783	Cret. (Cenom.)	8.9	8.9	0	8	1.95	17,29	0.96	11	2.0	16,24	−25.2	f
19	43-386-63-1 (142–144 cm)	31 11.2′ N 64 14.9′ W	937.2	4783	Cret. (Apt.)	4.1	3.3	7	5	1.49	23,31	0.46	5	3.9	16,24	−28.9	f
20	43-387-36-2 (145–150 cm)	32 19.2′ N 67 40.0′ W	558.3	5118	Cret. (Barr./Hauter.)	1.0	1.0	0	2.4	1.6	23,31	1.0	0.6	10.8	16	n.d.	f
21	43-387-37-2 (142–146 cm)	32 19.2′ N 67 40.0′ W	577.3	5118	Cret. (Barr./Hauter.)	3.9	3.9	0	n.d.	—	—	—	n.d.	—	—	n.d.	f
22	44-391A-21-4 (130–138 cm)	28 13.7′ N 75 36.9′ W	654.8	4963	Cret. (Alb.)	0.4	0.2	2	0.4	1.4	17,23,27	n.d.	0.04	3.6	16,24	n.d.	j
23	44-391C-7-2 (142–150 cm)	28 13.7′ N 75 36.9′ W	727.9	4963	Cret. (Alb.)	1.4	1.2	1	0.3	1.6	17,23	0.9	0.07	9.2	16,24	n.d.	j
24	44-391C-14-3 (142–150 cm)	28 13.7′ N 75 36.9′ W	1004.9	4963	Cret. (Barr.)	10.5	0.3	85	0.4	1.4	17,22,31	1.4	0.05	5.4	16,24	n.d.	j
25	44-391C-21-3 (138–146 cm)	28 13.7′ N 75 36.9′ W	1090.4	4963	Cret. (Barr./Berr.)	5.3	3.4	16	0.5	1.2	18,29	0.7	0.1	4.5	16,28	n.d.	j
26	47A-397A-51-4 (100–112 cm)	26 50.7′ N 15 10.8′ W	1444.0	2900	Cret. (Hauter.)	1.15	0.82	3	2.0	2.6	17,27	1.1	2.0	5.0	16,28	n.d.	k
27	478-398D-90-5 (15–30 cm)	40 57.6′ N 10 43.1′ W	1293.7	3890	Cret. (Alb.)	0.94	0.78	1	0.5	1.4	19,23,29	1.1	2.5	4.4	16,28	n.d.	k
28	478-398D-122-5 (132–142 cm)	40 57.6′ N 10 43.1′ W	1589.4	3890	Cret. (Apt.)	3.55	1.67	16	1.0	2.1	17,19,29	0.6	3.5	4.6	16,28	n.d.	k
29	48-402A-21-1 (40–44 cm)	47 52.5′ N 08 50.4′ W	327.4	2340	Cret. (Apt.)	4.64	1.73	24	17	2.7	19,23,33	1.4	28	3.2	16,28	n.d.	g
30	48-402A-30-1 (20–27 cm)	47 52.5′ N 08 50.4′ W	412.7	2340	Cret. (Apt.)	5.54	1.95	30	13	1.7	19,23,27	2.0	1.4	3.6	16,28	n.d.	g
31	62-466-29-2 (40–42 cm)	34 11.5′ N 179 15.3′ E	256.0	2665	Cret. (Alb.)	16.48	11.3	43	84	1.3	17	0.6	5.0	3.1	16,24	n.d.	l

* DSDP sample numbers are given as: Leg-Site-Core-Section (interval in section).

† Data supplied by G. Bode and S. M. White, DSDP, Scripps Institution of Oceanography, University of California, San Diego; Galbraith Laboratories, Knoxville, Tennessee and UCLA.

‡ Versus Chicago PDB standard.

§ Carbon preference index summed from C_{10} to C_{15}.

‖ The abundance maximum is underscored.

¶ Pristane-to-phytane ratio (Didyk et al., 1978).

** Reference to original data: (a) Simoneit et al., 1972; (b) Simoneit et al., 1973; (c) Simoneit et al., 1978; (d) Simoneit, 1977a; (e) Simoneit, 1978a; (f) Simoneit, 1979a; (g) Simoneit, 1979b; (h) Simoneit, 1981b; (i) Simoneit, 1980b; (j) Stuermer and Simoneit, 1978; (k) Simoneit and Mazurek, 1979a,b; (l) Simoneit, 1981a.

n.d.—not determined.

Table 2. Geochemical data for the kerogens of selected samples

Sample		Kerogen					ESR data‡		Reference§
Number (as in Table 1)	Designation	H/C*	N/C*	δ13C (‰)†	Sulfur (%)*	Vitrinite reflectance R_0 (mode %)	Spins/gram ×10^17	Line width (G)	
1	11-105-11-2	1.10	0.031	-26.7	n.d.	n.d.	9.1	6.12	a
2	14-138-6-3	n.d.	n.d.	-26.5	n.d.	n.d.	n.d.	n.d.	b
3	14-144-4-2	1.11	0.033	-26.7	n.d.	n.d.	1.9	7.14	b
4	14-144-4-3	1.05	0.026	-26.4	n.d.	n.d.	3.2	7.48	b
5	14-144A-5-1	1.33	0.026	-26.4	n.d.	n.d.	1.6	6.21	b
6	14-144A-6-1	1.21	0.029	-27.4	n.d.	n.d.	2.0	7.56	b
7	33-317A-16-2	1.20	0.006	-21.6 [-22.2]	12.7	0.25	6.2	6.38	g
8	36-330-4-2	1.17	0.002	-26.7	n.d.	n.d.	n.d.	n.d.	h
9	36-330-10-1	1.20	n.d.	-27.3	n.d.	n.d.	n.d.	n.d.	h
10	36-330-14-4	n.d.	n.d.	-24.1	n.d.	n.d.	n.d.	n.d.	h
11	40-364-24-1	1.41	0.021	-21.8 [-22.3]	10.2	n.d.	0.42	8.08	e
12	40-364-43-1	1.08	n.d.	-27.0	n.d.	n.d.	n.d.	n.d.	e
13	41-367-19-4	1.34	0.028	-28.9	13.7	n.d.	8.5	6.80	d
14	41-368-58-2	0.86	0.030	-25.0	n.d.	0.30	n.d.	n.d.	c
15	41-368-63-1	0.61	0.029	-27.2	6.5	0.80	n.d.	n.d.	c
16	41-368-63-2	1.43	0.037	-27.8	3.2	0.45	22.4	6.72	c
17	41-368-63-3	1.06	0.026	-27.4	8.5	0.35	27.5	7.23	c
18	43-386-43-3	0.59	0.024	-23.9	16.3	n.d.	1.8	6.97	c
19	43-386-63-1	n.d.	n.d.	-25.6	n.d.	n.d.	n.d.	n.d.	f
20	43-387-76-2	1.45	n.d.	-19.2	n.d.	n.d.	n.d.	n.d.	f
21	43-387-37-2	n.d.	n.d.	-26.6	n.d.	n.d.	n.d.	n.d.	f
22	44-391A-21-4	n.d.	n.d.	[-27.2]	n.d.	n.d.	n.d.	n.d.	i
23	44-391C-7-2	n.d.	n.d.	[-25.4]	n.d.	n.d.	n.d.	n.d.	i
24	44-391C-14-3	n.d.	n.d.	[-27.3]	n.d.	n.d.	n.d.	n.d.	i
25	44-391C-21-3	n.d.	n.d.	[n.r.]	n.d.	n.d.	n.d.	n.d.	i

* Data determined by Microanalytical Laboratory, Department of Chemistry, University of California at Berkeley and Galbraith Laboratories, Knoxville, Tennessee.
† Versus Chicago PDB standard, values in brackets are for humic substances.
‡ Corrected for ash content.
§ References to original data: (a) Simoneit et al., 1972; (b) Simoneit et al., 1973; (c) Simoneit et al. 1978; (d) Simoneit, 1977a; (e) Simoneit, 1978a; (f) Simoneit, 1979a; (g) Simoneit, 1981b; (h) Simoneit, 1980b; (i) Stuermer and Simoneit, 1978.
n.d.—not determined.
n.r.—none recoverable.

shown in Figure 2(A) and (D), respectively. Phytane is the most abundant hydrocarbon and Pr/Ph = 0.3 (porphyrin content = 6 ppm), indicating strong anoxic conditions of sedimentation (Didyk et al., 1978). The n-alkanes exhibit a bimodal distribution, with maxima at C_{17} and C_{31} and a high overall carbon preference index (CPI) (Simoneit, 1978b) of 2.9. The n-alkanes $> C_{23}$ are typical of higher plant wax, and the homologs $< C_{22}$ are typical of autochthonous production (Simoneit, 1975, 1978b, 1980a). The n-fatty acids also exhibit a bimodal distribution, with maxima at C_{16} and C_{24} and an overall CPI of 5.3. The homologs $> C_{20}$ are indicative of a higher plant origin, especially when considered with the wax alkanes, and those $< C_{19}$ are probably derived from autochthonous sources (Simoneit, 1975, 1978b). The concentration of phytanic acid exceeds that of pristanic acid (Figure 2(D)), further corroborating the euxinic paleoenvironment (Didyk et al., 1978). The lipids of the samples from Sites 138, 368, 397, 398 and 402 have similar distributions (cf. Table 1).

The samples from the south-eastern Atlantic (Site 364) are of mid- and lower-Cretaceous ages and the lipids are of varying origins. An example of the n-alkane and n-fatty acid distributions is shown in Figure 2(B) and (E), respectively. The n-alkanes exhibit a bimodal distribution, with maxima at C_{17} and C_{22}, and the n-fatty acids exhibit maxima at C_{16} and C_{22}. The dominance of n-heptadecane is attributable to a primary autochthonous origin from algae and the remaining homologs appear to be biodegraded residues from autochthonous marine and allochthonous terrigenous sources. The excess of both phytane and phytanic acid over pristane and pristanic acid, respectively, indicates strongly anaerobic sedimentary conditions.

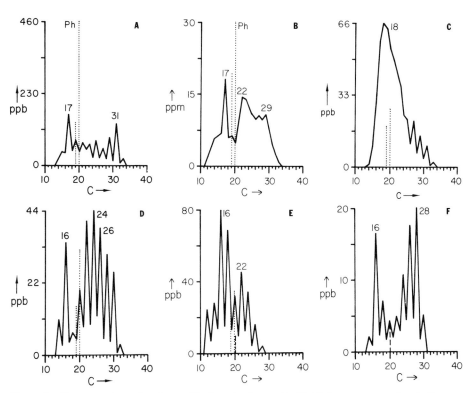

Figure 2.   Concentrations of n-alkanes (A–C) and n-fatty acids (D–F) for some examples: A, D: sample 41-367-19-4 (10–15 cm); B, E: sample 40-364-43-1 (130–133 cm); C, F: sample 44-391C-21-3 (138–146 cm); Isoprenoids (······) and diterpenoids (——) are also indicated.

The samples from the north-western Atlantic (Sites 105, 144, 386, 387 and 391) are representative of various Cretaceous epochs and the lipids have multiple origins. As was the case for the eastern areas, the sources of the n-alkanes and n-fatty acids were from autochthonous production and degradation, with some allochthonous terrigenous influx. The relative amount of each source category varies as can be seen in the distribution diagrams of three examples (Figures 2(C) and (F); 3(A), (B), (D) and (E). The lipids of the samples from Site 144 (e.g. Figure 3(A) and (D)) are predominantly of an autochthonous origin with primary production as the dominant source. The lipid yields were high compared to the other samples (Table 1).

The samples from the south-western Atlantic (Site 330) are of Lower Cretaceous and Upper Jurassic ages and the lipids have a mixed origin. As is the case for the south-eastern areas the origins of the n-alkanes and n-fatty acids reflect autochthonous production, biodegradation and allochthonous terrigenous influx. The relative amounts of each source category are variable.

In contrast, the shales from the Cretaceous Pacific Ocean appear to have a different genetic origin (Fischer *et al.*, 1977). Examples from the Manihiki Plateau (Site 317) and the Hess Rise (Site 466) were analyzed. The sample from Site 317 is a paper shale which occurs as a 4 cm horizon above volcanigenic sediments and under carbonates. It represents probably a lagoonal environment during a brief period of the Cretaceous. The n-alkanes exhibited a trimodal distribution with maxima at C_{18}, C_{22} and C_{31} and the n-fatty acids were also trimodal with maxima at C_{16}, C_{24} and C_{28}. The n-alkanes $> C_{26}$ and the n-fatty acids $> C_{21}$ are derived from higher plant wax, whereas, the homologs below these molecular weights are due to a marine

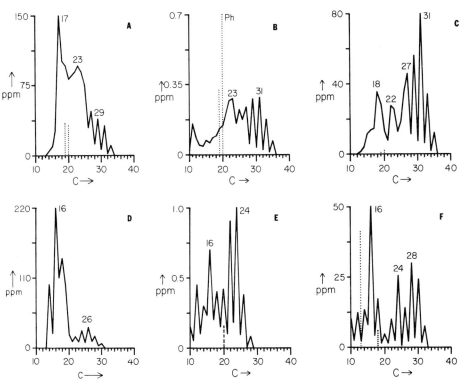

Figure 3.   Concentration of n-alkanes (A–C) and n-fatty acids (D–F) for some examples: A, D sample 14-144A-6-1 (100–101 cm); B, E: sample 43-386-63-1 (142–144 cm); C, F: sample 33-317A-16-2 (131–134 cm).

autochthonous origin (Simoneit, 1978b, 1981b). Only trace amounts of phytane and pristane are present and the isoprenoidal ketones consist of predominantly 6,10-dimethylundecan-2-one. No porphyrin pigments were detected and perylene was also absent. The claystone from Site 466 contains lipids of a primarily autochthonous marine origin, where the n-alkanes exhibited a distribution maximum at C_{17} and the n-fatty acids had a dominant maximum at C_{16}. The paleoenvironmental conditions of sedimentation were anoxic with a high autochthonous productivity (Simoneit, 1981a).

3.2. Molecular markers

Various other molecular markers which are attributable to specific biogenic sources were identified in the lipids of these sediments.

The isoprenoidal ketones, mainly 6,10,14-trimethylpentadecan-2-one (Structure I) and to a lesser extent 6,10-dimethylundecan-2-one, are present in the lipids of most of these samples. These compounds indicate microbial alteration of the lipids, especially of phytol and its diagenetic products (Ikan et al., 1973; de Leeuw et al., 1977; Simoneit, 1973).

Diterpenoids were identified in most of the samples from the Atlantic, but no analogs were detected in the samples from the Pacific (Sites 317 and 466). Dehydroabietic acid (II) was confirmed in samples from Sites 330, 364, 367, 368, 387, 391, 397, 398 and 402. Various diterpenoidal hydrocarbons, mainly dehydroabietin (III), retene (IV), iosene (V), and simonellite (VI) were identified in some of these samples (e.g., Simoneit, 1979a,b; 1978a; 1977a). Diterpenoids have been characterized in many DSDP sediments of younger ages (Simoneit, 1977b).

Two sesquiterpenoids are dominant constituents of the hydrocarbon fraction from the Site 317 sample and minor occurrences in some of the other samples. Cadalene (VII) and a tetrahydrocadalene (VIII) were identified and they may be of either marine algal or terrestrial origin.

Steranes, C_nH_{2n-6} (IX) and ster-4-enes (also ster-2-enes), C_nH_{2n-8} (X) ranging from $n=27$ to 29 were identified in samples from Sites 105, 144, 330, 364, 367, 368, 386, 387, 398 and 466 in amounts ranging from ppm to traces. The relative amounts of saturated and unsaturated compounds are also variable. Backbone rearranged sterenes (diasterenes) were identified in samples from Sites 364, 386 and 398. The diasterenes, C_nH_{2n-8} (XI), ranged from $n=27$ to 29. These steroidal compounds appear to be of a predominantly autochthonous origin, and they are not detectable in the sample from Site 317.

Triterpenoidal compounds were identified in most samples from all sites and they consisted predominantly of hydrocarbons and acids, with only traces of ketones. The carbon skeletons found were mainly the hopane series, some moretanes and several unknowns. Trisnorhopane (XII), norhopane (XIII), hopane (XIV), and extended hopanes (XV), with predominantly the $17\beta(H)$ stereochemistry in most cases, were present in the geologically immature samples (e.g., Sites 364 and 386, Simoneit, 1978a, 1979a). Hop-17(21)-ene (XVI) was also present as a significant component in immature samples. Some of the samples contained predominantly $17\alpha(H)$-hopanes and many had both $17\alpha(H)$ and $17\beta(H)$ analogs. Some examples of the relative distributions of triterpenoids, based on the m/z 191 peak intensity in the GC/MS data, are given in Figure 4. For sample 364-24-1, the dominant species were trisnorhopane, hopane and homohopane with the $17\beta(H)$ stereochemistry, with lesser amounts of $17\alpha(H)$ stereomers and mono-enes (Figure 4(A)). In the case of sample 386-63-1, the dominant homologs were again primarily the $17\beta(H)$-hopanes (Figure 4(B)). For comparison with these distributions, two examples of recent sediments are also presented. The sample from Mangrove Lake, Bermuda, consists

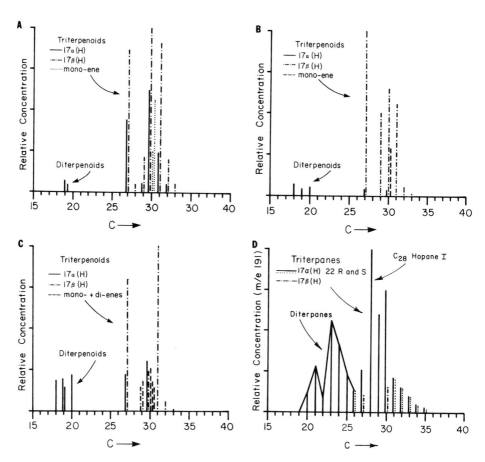

Figure 4.   Relative concentrations of triterpenoids and diterpenoids in sediment lipids; (A) sample 40-364-24-1 (0–10 cm); (B) sample 43–386–63–1 (142–144 cm); (C) Mangrove Lake, Bermuda (recent, sample ML 71-2-31, Hatcher *et al.*, 1977); (D) Southern California Bight (recent, BLM 386-30, cm, Simoneit and Kaplan, 1980).

mainly of algal detritus in the form of a sapropel (Hatcher *et al.*, 1977). The dominant triterpenoids consisted of trisnorhopane and homohopane with lesser amounts of norhopane and hopane, all with the $17\beta(\mathrm{H})$ stereochemistry (Figure 4(C)). The minor components were other $17\alpha(\mathrm{H})$ and $17\beta(\mathrm{H})$ homologs and unsaturated analogs. The sample from the Southern California Bight consists of hemipelagic detritus with a significant terrigenous component (Simoneit & Kaplan, 1980). The triterpanes (Figure 4(D)) were comprised primarily of the $17\alpha(\mathrm{H})$-hopane series ranging from C_{27} to C_{35}, with the homologs $>C_{31}$ as 1 : 1 mixtures of the 22R and 22S stereomers. These compounds are indicative of geologically mature lipids and were derived from paleoseepage of petroleum (Dastillung & Albrecht, 1976; Simoneit & Kaplan, 1980).

Most samples contained triterpenoidal acids, $C_nH_{2n-10}O_2$ (XVII), ranging from $n = 31$ to 35 and $n = 32$ as the major homolog. These hopanoic acids usually have the $17\beta(\mathrm{H}) > 17\alpha(\mathrm{H})$ stereochemistry, where the $17\beta(\mathrm{H})$ configuration is of a recent biogenic (microbial) origin (Ourisson *et al.*, 1979).

The appearance of the $17\alpha(\mathrm{H})$-hopane stereochemistry, which is thermodynamically more stable, with greater depth of burial has been used as a gauge of petrogenic maturity of the organic matter (Dastillung & Albrecht, 1976). The presence of large amounts of 17β(H)-hopanes indicates that these Cretaceous sediments are extremely immature and have experienced limited elevated temperatures. The sample from Site 317 contains essentially only 17β(H)-homohopane (XV, R=CH₃) and no triterpenoidal acids or ketones. These triterpenoids appear to be molecular markers derived predominantly from autochthonous microbial sources in the marine environment (Ourisson et al., 1979).

Partially and fully aromatized triterpenoids (Spyckerelle et al., 1977; Greiner et al., 1976, 1977), which appear to be diagenetic products, were identified in samples from Sites 364, 367, 402 and 466 (Simoneit, 1977a, 1978a, 1979b, 1981a). The monoaromatic 14,18-bisnormethyladianta-13,15,17-triene (XVIII) was characterized in all samples. Samples 364-24-1 and 466-24-2 contained the complete series of the aromatic triterpenoids, where the additional analogs are comprised of 8,14,18-trisnormethyladianta-8,11,13,15,17-pentaene (XIX), (3'-ethylcyclopenteno-7,8)1,1-dimethyltetrahydro(1,2,3,4)-chrysene (XX), and (3'-ethylcyclopenteno-7,8)1-methylchrysene (XXI). A higher homologous series with one additional methyl substituent in probably the A or B ring was also present in small quantities.

Perylene (XXII) was present in most of the samples. The highest concentrations were found for the samples with lipids of a terrigenous nature deposited in euxinic paleoenvironments (e.g., sample 386-63-1, versus traces only in 386-43-3, Simoneit, 1979a). Perylene may be a terrigenous marker (Aizenshtat, 1973), since it was encountered mainly in sediments with organic matter of predominantly such an origin (Simoneit, 1980a). The strongly euxinic paleoenvironments of many of these samples may have enhanced the formation and preservation of perylene (Didyk et al., 1978).

The stable carbon isotope results for the total lipid fractions are given in Table 1. The $\delta^{13}C$ values in the range of -24 to $-29‰$ corroborate the terrigenous origin of the lipids (Degens, 1969; Kaplan, 1975) and correlate well with the molecular markers. The $\delta^{13}C$ value of the lipids from Site 317 is of a more marine character (Degens, 1969; Kaplan, 1975).

3.3 Kerogens

The geochemical data for the kerogens and some humates are given in Table 2. Most of the $\delta^{13}C$ values lie in the range -24 to $-29‰$ which can be considered as typical of a terrigenous origin for immature sediments (Degens, 1969; Kaplan, 1975). The kerogens with $\delta^{13}C$ values of a more marine character ($> -24‰$) were from samples of 317A-16-2, 364-24-1 and 387-36-2. Essentially, all $\delta^{13}C$ values of the kerogens were isotopically heavier than those for the corresponding lipids.

A correlation plot of the atomic H/C versus the $\delta^{13}C$ values of kerogens and humic substances from both Recent (Stuermer et al., 1978) and Cretaceous environments is shown in Figure 5. The data indicate a clustering, which reflects the genetic origin, i.e. terrigenous, marine or a mixture of both (Simoneit, 1980c, 1981b). Samples with an inferred terrigenous origin aggregate in the more aromatic and isotopically lighter region and the autochthonous marine samples are found in the more aliphatic and isotopically heavier region.

The high sulfur contents of the kerogens (Table 2) also support the anaerobic sedimentary environment (Didyk et al., 1978). The limited vitrinite reflectance data indicate that the kerogens are geologically immature. The R_0 values of <0.5 are representative of thermally unaltered kerogens (Peters et al., 1979).

ESR spectroscopy has been used to distinguish sources of kerogenous materials (Ho, 1977; Stuermer et al., 1978). Humic-type (terrigenous) kerogens have a higher spin density than the sapropelic-type (algal). The data for these Cretaceous kerogens with similar data for kerogens from recent sources are plotted versus H/C in Figure

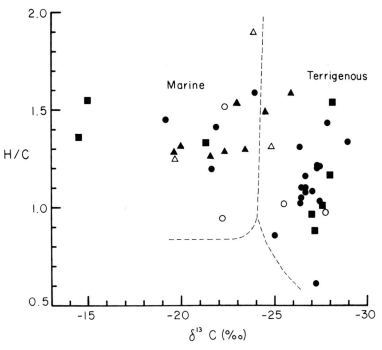

Figure 5. Correlation diagram of the atomic H/C versus δ^{13} C (per mil versus PDB) for some kerogens and humic substances (symbols: ▲=Recent kerogen; ●=Cretaceous kerogen; ■=Recent kerogen, Stuermer *et al.*, 1978; △=Recent and ○=Cretaceous humic substances).

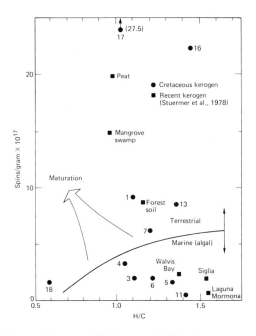

Figure 6. Correlation diagram of the atomic H/C versus the ESR spin concentration per gram of kerogen from Cretaceous and Recent (Stuermer *et al.*, 1978) samples. Sample numbers refer to Table 1.

6. The samples corresponding to vascular plant sources have high spin concentrations and lower H/C values, reflecting their aromatic character. Whereas, the marine (algal) samples have low spin concentrations and higher H/C values.

The inferred origins of the bulk of the organic matter in the various Cretaceous paleoenvironments of the Atlantic and Pacific Oceans are summarized in Table 3. In general, during the Early Cretaceous (Neocomian to Cenomanian) the origin of the organic matter was predominantly terrigenous in both the North and South Proto-Atlantic. This is illustrated by Figure 7(a) where the major sources of the organic matter and the dominant oceanic currents are indicated. The terrestrial contribution of organic detritus appears to have been due to the widespread shallow epicontinental seas (e.g. Cox, 1973), produced by sea-level changes associated with vertical continental motions which flooded continental areas (Schlanger & Jenkyns, 1976; Bond, 1978). Thus, the high influx of terrigenous detritus from primary production and secondary erosion combined with a reduced supply of cold oxygenated bottom waters and resultant oxygen-minimum layers (Schlanger & Jenkyns, 1976; Thiede & van Andel, 1977) favored the preservation of these organic carbon-rich sediments. For the Late Cretaceous (Turonian to Maestrichtian) the data here are still preliminary, but it appears that marine autochthonous sedimentation of organic matter became more important. This is indicated in Figure 7(b). The data for the sample from Site 317 indicate that it was deposited by high autochthonous productivity in a shallow lagoonal environment in the mid-Pacific Ocean. The data for the sample from Site 466 indicate high autochthonous marine productivity on the slopes of the Hess Rise.

Table 3. Inferred origins of the bulk organic matter in the various Cretaceous paleoenvironments of the Atlantic and Pacific Oceans

		Inferred origin	
Site	Cretaceous	Lipids*	Kerogen (humate)[1]
North-eastern Atlantic			
138	Early (lower)	Terrigenous/marine	Terrigenous
367	Early	Terrigenous/marine	Terrigenous
368	Early	Terrigenous/marine	Terrigenous
397	Early	Terrigenous	—
398	Early	Terrigenous/marine	—
398	Late (upper)	Terrigenous/marine	—
402	Early	Terrigenous/marine	—
North-western Atlantic			
105	Early	Marine	Terrigenous
144	Late	Terrigenous/marine	Terrigenous
386	Early	Terrigenous/marine	Terrigenous/marine
387	Early	Marine	Marine
391	Early	Terrigenous/marine	(Terrigenous)
South-eastern Atlantic			
364	Early	Terrigenous/marine	Terrigenous
364	Late	Terrigenous/marine	Marine (marine)
South-western Atlantic			
330	Jurassic	Terrigenous/marine	Marine
330	Early	Terrigenous/marine	Terrigenous
Pacific			
317	Early	Terrigenous/marine	Marine (marine)
466	Early	Marine	Marine (terrigenous?)

* The dominant origin is underscored.

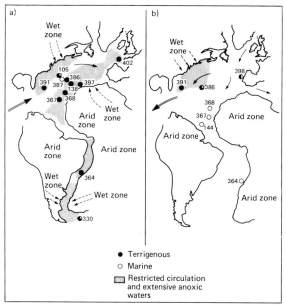

Figure 7. Paleogeographic reconstructions of the Atlantic Ocean during the Cretaceous, indicating the nature of the sedimentary organic matter, the restricted oceanic circulation with allied anoxic conditions and the terrestrial arid and wet areas: (a) Early (Lower) Cretaceous and (b) Late (Upper) Cretaceous.

4. Conclusions

During Cretaceous time, the basins and continental slopes of the Proto-Atlantic underwent various sedimentary regimes. The input of organic carbon to these sediments resulted from the unusually high supply of pelagic detritus, the massive and rapid supply of terrigenous plant matter, and/or the redeposition of ancient kerogen (some coal). These conclusions are supported by the lipid data and by the results of the kerogen analyses. The organic carbon-rich black shales and some of the claystones formed in areas where oxygen-minimum layers developed in the upper mid-water masses or where the conditions were completely euxinic (Fischer et al., 1977). The presence of porphyrins and sulfur and of $Pr/Pr < 1$ corroborate such anoxic depositional environments (Didyk et al., 1978). The lipid matter of many samples is catagenetically immature, indicating a mild geothermal history for these sites.

The kerogens and humates of most samples from the Atlantic were derived from terrigenous sources of primary production and recycling, especially in the Early Cretaceous. For the limited number of samples that have been examined from the Late Cretaceous, the bulk of the organic matter appears to be derived from autochthonous marine production (deeper waters).

The lipids of a Cretaceous paper shale from Site 317 in the mid-Pacific Ocean appear to be derived from predominantly autochthonous algal production with an influx of higher plant wax. The kerogen is typically marine. These data indicate that this paleoenvironment probably consisted of a highly productive lagoon on an uplifted rise or atoll, unlike the paleoenvironments of the early Atlantic Ocean. The organic matter of the sample from Site 466 was probably preserved under a mid-water oxygen minimum zone in an area of high marine productivity on the Hess Rise (e.g., Demaison & Moore, 1980).

Appendix

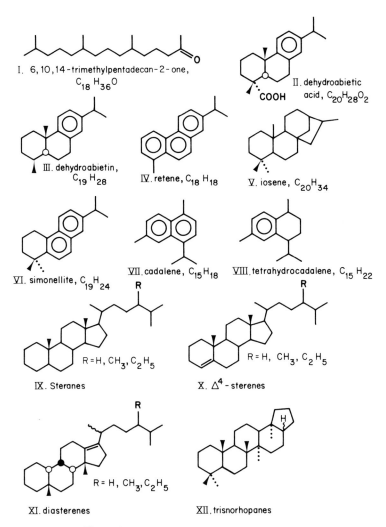

I. 6,10,14-trimethylpentadecan-2-one, $C_{18}H_{36}O$

II. dehydroabietic acid, $C_{20}H_{28}O_2$

III. dehydroabietin, $C_{19}H_{28}$

IV. retene, $C_{18}H_{18}$

V. iosene, $C_{20}H_{34}$

VI. simonellite, $C_{19}H_{24}$

VII. cadalene, $C_{15}H_{18}$

VIII. tetrahydrocadalene, $C_{15}H_{22}$

IX. Steranes   R = H, CH_3, C_2H_5

X. Δ^4-sterenes   R = H, CH_3, C_2H_5

XI. diasterenes   R = H, CH_3, C_2H_5

XII. trisnorhopanes

Figure A1.   Chemical structures cited in the text.

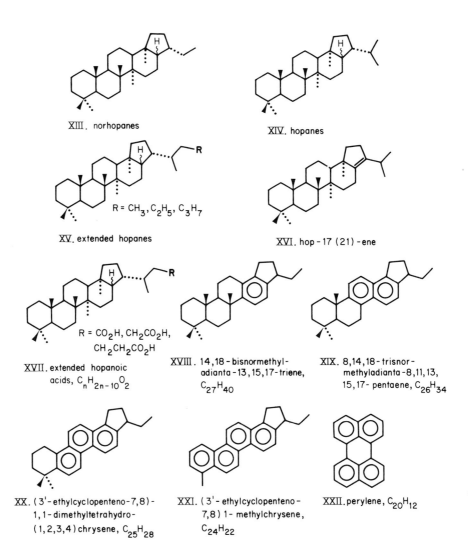

XIII. norhopanes

XIV. hopanes

XV. extended hopanes

R = CH$_3$, C$_2$H$_5$, C$_3$H$_7$

XVI. hop - 17 (21) - ene

XVII. extended hopanoic
acids, C$_n$H$_{2n-10}$O$_2$

R = CO$_2$H, CH$_2$CO$_2$H,
CH$_2$CH$_2$CO$_2$H

XVIII. 14,18 - bisnormethyl-
adianta - 13,15,17 - triene,
C$_{27}$H$_{40}$

XIX. 8,14,18 - trisnor-
methyladianta - 8,11,13,
15,17 - pentaene, C$_{26}$H$_{34}$

XX. (3' - ethylcyclopenteno - 7,8) -
1,1 - dimethyltetrahydro -
(1,2,3,4) chrysene, C$_{25}$H$_{28}$

XXI. (3' - ethylcyclopenteno -
7,8) 1 - methylchrysene,
C$_{24}$H$_{22}$

XXII. perylene, C$_{20}$H$_{12}$

Figure A1 (cont.).   Chemical structures cited in the text.

Acknowledgments

We thank the National Science Foundation via the Deep Sea Drilling Project for making the DSDP Core material available for study and Ms M. A. Mazurek and Mr E. Ruth for technical assistance. Contribution No. 2068 from the Institute of Geophysics and Planetary Physics, University of California at Los Angeles.

References

Aizenshtat, Z. 1973. Perylene and its geochemical significance. *Geochimica et Cosmochimica Acta* **37**, 559–567.

Arthur, M. A. & Schlanger, S. O. 1977. Middle Cretaceous 'ocean anoxic events' as causal factors in development of giant oil fields. *American Association of Petroleum Geologists Bulletin* **61**, 762.

Arthur, M. A. & Schlanger, S.O. 1979. Cretaceous 'oceanic anoxic events' as causal factors in development of reef-reservoired giant oil fields. *American Association of Petroleum Geologists, Bulletin* **63**, 870–885.

Bogolyubova, I. & Timofeev, P. P. 1978. Composition of organic matter in 'Black Shales' of the Cape Verde Basin (Eastern Atlantic) and their potential as petroleum. *Lithology and Mineral Resources* **13**, 519–530.

Bond, G. 1978. Speculations of real sea-level changes and vertical motions of continents at selected times in the Cretaceous and Tertiary Periods. *Geology* **6**, 247–250.

Cox, A. 1973. *Plate Tectonics and Geomagnetic Reversals*. San Francisco: Freeman. p. 116.

Dastillung, M. & Albrecht, P. 1976. Molecular test for oil pollution in surface sediments. *Mar. Poll. Bull.* **7**, 13–15.

Degens, E. T. & Ross, D. A. 1974. *The Black Sea—Geology, Chemistry, Biology*, AAPG mem. **20**, 633pp.

Degens, E. T. & Stoffers, P. 1976. Stratified waters as a key to the past. *Nature* **263**, 22–27.

Degens, E. T. 1969. Biogeochemistry of stable carbon isotopes. *Organic Geochemistry, Methods and Results* (Eds G. Eglinton & M. T. J. Murphy), Berlin: Springer-Verlag, pp. 304–329.

de Leeuw, J. W., Simoneit, B. R., Boon, J. J., Rijpstra, W. I. C., de Lange, F., van der Leeden, J. C. W., Correia, V. A., Burlingame, A. L. & Schenck, P. A. 1977. Phytol-derived compounds in the geosphere. *Advances in Organic Geochemistry 1975* (Eds R. Campos & J. Goni), Enadimsa, Madrid, 61–79.

Demaison, G. J. & Moore, G. T. 1980. Anoxic environments and oil source bed genesis. *Organic Geochemistry* **2**, 9–31.

Didyk, B. M., Simoneit, B. R. T., Brassell, S. C. & Eglinton, G. 1978. Organic geochemical indicators of paleoenvironmental conditions of sedimentation. *Nature* **272**, 216–222.

Dow, W. G. 1977. Petroleum source beds on Continental slopes and rises. In *Geology of Continental Margins*, AAPG Cont. Education Ser. **5**, D1–D37.

Emery, K. O. 1960. *The Sea Off Southern California*. New York: John Wiley and Sons, 366 pp.

Fischer, A. G., Arthur, M. A., Herb, R. & Silva, I. P. 1977. Middle Cretaceous events. *Geotimes* **22** (4), 18–19.

Gartner, S. 1979. Paleooceanographic catastrophe at the end of the Cretaceous, Abstr., *EOS*, **60**, 854.

Greiner, A. C., Spyckerelle, C., Albrecht, P. & Ourisson, G. 1977. Hydrocarbures aromatiques d'origine geologique. V. Derivés mono- et di-aromatiques du hopane. *Journal of Chemical Research* (M), 3829–3871.

Greiner, A. C., Spyckerelle, C. and Albrecht, P. 1976. Aromatic hydrocarbons from geological sources— I. New naturally-occurring phenanthrene and chrysene derivatives. *Tetrahedron* **32**, 257–260.

Hatcher, P. G., Simoneit, B. R. T. and Gerchakov, S. M. 1977. The organic geochemistry of a Recent sapropelic environment: Mangrove Lake, Bermuda. In *Advances in Organic Geochemistry 1975* (Eds R. Campos & J. Goni), Enadimsa, Madrid, 469–484.

Ho, T. T. Y. 1977. Geological and geochemical factors controlling electron spin resonance signals in kerogen. *ASCOPE/CCOP Seminar on Generation and Maturation of Hydrocarbons in Sedimentary Basins*, Manilla, UN ESCAT/CCOP Techn. Publ. No. 6, pp. 54–80.

Ikan, R., Baedecker, M. J. & Kaplan, I. R. 1973. C_{18}-isoprenoid ketone in Recent marine sediment. *Nature* **244**, 154–155.

Kaplan, I. R. 1975. Stable isotopes as a guide to biogeochemical processes. *Proceedings of the Royal Society of London Series B* **189**, 183–211.

Kaplan, I. R., Smith, J. W. & Ruth, E. 1970. Carbon and sulfur concentration and isotopic composition in Apollo 11 lunar samples. *Proceedings of Apollo 11 Lunar Science Conference—Geochimica et Cosmochimica Acta, Supplement 1*, Vol. 2, Pergamon, 1317–1329.

Ourisson, G., Albrecht, P. & Rohmer, M. 1979. The Hopanoids, palaeochemistry and biochemistry of a group of natural products. *Pure & Applied Chemistry* **51**, 709–729.

Peters, K. E., Simoneit, B. R. T., Brenner, S. & Kaplan, I. R. 1979. Vitrinite reflectance—temperature determinations for intruded Cretaceous black shale in Eastern Atlantic. *Symp. Low Temperature Metamorphism of Kerogen and Clay Minerals* (Ed. D. F. Oltz), Pac. Sect., Soc. Econ. Paleontol. Mineral., Los Angeles, pp. 53–58.

Philippi, G. T. 1974. The influence of marine and terrestrial source material on the composition of petroleum. *Geochimica et Cosmochimica Acta* **38**, 947–966.

Ryan, B. F. & Cita, M. B. 1977. Ignorance concerning episodes of ocean-wide stagnation. *Marine Geology* **23**, 197–215.

Schlanger, S. O. & Jenkyns, H. C. 1976. Cretaceous oceanic anoxic events: Causes and consequences. *Geologie en Mijnbouw* **55**, 179–184.

Simoneit, B. R. 1973. Identification of isoprenoidal ketones in Deep Sea Drilling Project core samples and their geochemical significance. *Initial Reports of the Deep Sea Drilling Project* **21**, 909–923.

Simoneit, B. R. T. 1975. Sources of organic matter in oceanic sediments. *Ph.D. Thesis*, University of Bristol, England, 300 pp.

Simoneit, B. R. T. 1977a. Leg 41 sediment lipids—Search for eolian organic matter in Recent samples and examination of a black shale. *Initial Reports of the Deep Sea Drilling Project* **41**, 855–858.

Simoneit, B. R. T. 1977b. Diterpenoid compounds and other lipids in deep-sea sediments and their geochemical significance. *Geochimica et Cosmochimica Acta* **41**, 463–476.

Simoneit, B. R. T. 1978a. Lipid analyses of sediments from site 40-364 in the Angola Basin. *Initial Reports of the Deep Sea Drilling Project* **40**, 659–662.

Simoneit, B. R. T. 1978b. The organic chemistry of marine sediments. *Chemical Oceanography* (Eds J. P. Riley & R. Chester), 2nd edn, Vol. 7. New York: Academic Press.Pp. 233–311.

Simoneit, B. R. T. 1979a. Organic geochemistry of the shales from the Western Proto-North Atlantic, DSDP Leg 43. *Initial Reports of the Deep Sea Drilling Project* **43**, 643–649.

Simoneit, B. R. T. 1979b. Organic geochemistry of Cretaceous black shales from the Bay of Biscay, site 402 and of Eocene mudstone from the Rockall Plateau, site 404. *Initial Reports of the Deep Sea Drilling Project* **48**, 935–941.

Simoneit, B. R. T. 1980a. Terrigenous and marine organic markers and their input to marine sediments. *Proceedings of Symposium on Organic Geochemistry of Deep Sea Drilling Project Sediments* (Ed. E. W. Baker), Princeton: Science Press, in press.

Simoneit, B. R. T. 1980b. Organic geochemistry of Mesozoic sediments from Deep Sea Drilling Project Site 330, Falkland Plateau. *Initial Reports of the Deep Sea Drilling Project* **50**, 637–642.

Simoneit, B. R. T. 1980c. Utility of molecular markers and stable isotope compositions in the evaluation of sources and diagenesis of organic matter in the geosphere. *The Impact of the Treibs' Porphyrin Concept on the Modern Organic Geochemistry* (Ed. A. Prashnowsky), Universität Würzburg, pp. 133–158.

Simoneit, B. R. T. 1981a. Organic geochemistry of Albian sediment from the Hess Rise, DSDP-IPOD Leg 62, *Initial Reports of the Deep Sea Drilling Project* **62**, 939–942.

Simoneit, B. R. T. 1981b. Organic geochemistry of the shales and sapropels of the Cretaceous Atlantic. *Organic Geochemistry of Oil, Gas and Organic Matter of the Precambrian*, (Eds A. V. Sidorenko and N. A. Eremenko), USSR Acad. Sciences, Moscow, pp. 138–151.

Simoneit, B. R. T., Brenner, S., Peters, K. E. & Kaplan, I. R. 1981. Thermal alteration of Cretaceous black shale by diabase intrusions in the Eastern Atlantic. II: Effects on bitumen and kerogen. *Geochimica et Cosmochimica Acta*, **45**, 1581–1602.

Simoneit, B. R. T., Brenner, S., Peters, K. E. and Kaplan, I. R. 1978. Thermal alteration of Cretaceous black shale by basaltic intrusions in the Eastern Atlantic, *Nature* **273**, 501–504.

Simoneit, B. R. T. & Kaplan, I. R. 1980. Triterpenoids as molecular indicators of paleoseepage in Recent sediments of the Southern California Bight. *Marine Environment Research* **3**, 113–128.

Simoneit, B. R. T. and Mazurek, M. A. 1979a. Search for eolian lipids in the Pleistocene off Cape Bojador and lipid geochemistry of a Cretaceous mudstone, DSDP/IPOD, Leg 47A. In: *Initial Reports of the Deep Sea Drilling Project*, Vol. 47, Part I, (Ryan, W. B. F., Rad, U. von, *et al.*), U.S. Government Printing Office, Washington, D.C., pp. 541–545.

Simoneit, B. R. T. and Mazurek, M. A. 1979b. Lipid geochemistry of Cretaceous sediments from Vigo Seamount, DSDP/IPOD Leg 47B. In: *Initial Reports of the Deep Sea Drilling Project*, Vol. 47, Part II, (Sibuet, J. C., Ryan, W. B. F., *et al.*), U.S. Government Printing Office, Washington, D.C., pp. 565–570.

Simoneit, B. R. T., Scott, E. S. & Burlingame, A. L. 1973. Preliminary organic analysis of DSDP cores, Leg 14, Atlantic Ocean. *Initial Reports of the Deep Sea Drilling Project* **16**, 575–600.

Simoneit, B. R. T., Scott, E. S., Howells, W. G. & Burlingame, A. L. 1972. Preliminary organic analyses of the Deep Sea Drilling Project Cores, Leg 11. *Initial Reports of the Deep Sea Drilling Project* **11**, 1013–1045.

Spyckerelle, C., Greiner, A. C., Albrecht, P. & Ourisson, G. 1977. Hydrocarbures aromatiques d'origine geologique. III. Un tetrahydrochrysène, dérivé de triterpènes, dans les sediments recents et anciens: 3,3,7-trimethyl-1,2,3,4-tétrahydrochrysène. *Journal of Chemical Research* (M) 3746–3777.

Stuermer, D. H. and Simoneit, B. R. T. 1978. Varying sources for the lipids and humic substances in

Site 44-391, Blake Bahama Basin. In: *Initial Reports of the Deep Sea Drilling Project*, Vol. 44, (Benson, W. E., Sheridan, R. E., *et al.*), U.S. Government Printing Office, Washington, D.C., pp. 587–591.

Stuermer, D. H., Peters, K. E. & Kaplan, I. R. 1978. Source indicators of humic substances and proto-kerogen. Stable isotope ratios, elemental compositions and electron spin resonance spectra. *Geochimica et Cosmochimica Acta* **42**, 989–997.

Thiede, J. & van Andel, T. H. 1977. The paleoenvironment of anaerobic sediments in the late Mesozoic South Atlantic Ocean. *Earth and Planetary Science Letters* **33**, 301–309.

Tissot, B., Demaison, G. J., Masson, P. Delteil, J. R. & Combaz, A. 1980. Paleoenvironment and petroleum potential of Mid-Cretaceous black shales in Atlantic Basins, Abstr., *American Association of Petroleum Geologists, Bulletin*, **64**, 2051–2063.

Tissot, B., Deroo, G. & Herbin, J. P. 1979. Organic matter in Cretaceous Sediments of the North Atlantic: contribution to Sedimentology and paleogeography. *Deep Drilling Results in the Atlantic Ocean: Continental margins and Paleoenvironment* (Eds M. Talwani, W. Hay & W. B. F. Ryan), Maurice Ewing Series 3. Washington: American Geophysical Union. Pp. 362–374.

Tissot, B., Durand, B., Espitalié, J. & Combaz, A., 1974. Influence of nature and diagenesis of organic matter in formation of petroleum. *American Association of Petroleum Geologists, Bulletin* **58**, 499–506.

9. Nature and Origin of Late Neogene Mediterranean Sapropels

M. B. Cita

Institute of Paleontology, University of Milan

D. Grignani

AGIP-SGEL, San Donato, Milan

Oxygen-deficient environments resulting in intermittent stagnant episodes characterized the eastern Mediterranean basins after the closure of marine communications with the Indian Ocean, in mid-Miocene time. Oxygen depletion, extending into the western basins, increased drastically prior to the Messinian salinity crisis. Anaerobic conditions occurred in basins subjected to desiccation.

Sapropel deposition was limited to the eastern Mediterranean after the termination of the Messinian crisis, in the Pliocene and Quaternary. Dozens of individual sapropels are recorded in pelagic sections: the most prominent ones can be correlated with definite times of transgression. Organic carbon contents, mainly of marine origin, range up to 16 %.

Late Pleistocene sapropels, which are the best known, originated in different ways. Some of these sapropels seem to be modulated according to Milankovic's precession cycles, others according to cycles related to the inclination of the Earth's axis. Isotopic composition of oxygen in Mediterranean sapropels is markedly more negative than in correlative levels in the open ocean. Isotopic lightening and peaks of abundance of planktonic foraminiferal species tolerant to low salinities are attributed to the dilution of surficial water masses induced by glacial runoff from large continental ice sheets. Investigations on the quantitative composition of benthic foraminiferal assemblages record drastic changes related to basin stagnation.

1. Introduction

The Mediterranean is an enclosed basin, entirely surrounded by land masses. The connections with the Atlantic Ocean through the Gibraltar Straits, which are some 350 m deep, account for the present lack of an oceanic-type thermo-haline circulation at depth.

After the Alpine orogeny, paleogeographic conditions leading to deposition and preservation of organic-rich layers first developed in Serravallian times, as proved by deep-sea drilling (Ryan, Hsü *et al.*, 1973; Hsü, Montadert *et al.*, 1978).

Sapropel-bearing sequences are described in the land record since the Langhian, and are interpreted as a response to fragmentation of the Mediterranean by post-orogenic block faulting (Meulenkamp *et al.*, 1979); however, there is no documentation concerning the organic carbon content of these sequences, so they will not be considered here.

The intentions of this paper are (1) to provide a general overview of Late Neogene sapropel deposition in the Mediterranean, essentially based on data from the deep sea, (2) to provide new analytical data on pyrolysis assays and kerogen composition on sapropels of Miocene, Pliocene and Pleistocene age from the Deep Sea Drilling Project, (3) to provide new information on kerogen composition and mineralogical composition (by X-ray diffractometry) on sapropels of Holocene and Late

Pleistocene age from the Mediterranean and Calabrian Ridges, and (4) to discuss various aspects of sapropel deposition in the youngest part of the stratigraphic column, which is by far the best known.

Older sapropels are too imperfectly known to permit the formulation of well-founded models on their origin, but a comparison with the younger sapropels may be attempted.

2. Distribution and characters of Miocene organic carbon-rich facies

Organic carbon-rich layers were recovered repeatedly in the Serravallian and Tortonian of the eastern Mediterranean basis (DSDP Sites 126, 375) whereas they are conspicuously absent in the correlative stratigraphic interval in the western Mediterranean Site 372.

Figure 1 and Table 1 contain information on the location of the drill sites, piston-cores and other localities discussed in this paper. Figure 2 shows schematically the Neogene stratigraphic framework of the Mediterranean Neogene and the approximate position of the organic carbon-rich intervals in the western and eastern Mediterranean basins.

Table 2 contains data on organic carbon content, pyrolysis assays, kerogen composition and pyrochromatographic analysis of 16 sapropel samples from DSDP Leg 42A. Four sapropels, all from DSDP Site 375 in the Levantine Basin are Miocene in age. The organic carbon content of these layers is up to 6%. The HK/CO ratio is very high for the samples containing high percentages of organic carbon, and includes the highest value measured. Kerogen composition ranges from 70–90% amorphous organic matter, 0–15% fusinite, 0–5% herbaceous continental

Table 1.   Coordinates and water depth of piston-cores and drillsites discussed in this paper

Piston-cores	Drillsites	Latitude N	Longitude E	Water depth (m)
Alb 189	DSDP 124	33°54'	28°29'	2664
RC9-181	DSDP 125	33°25'	25°01'	2286
RC9-185	DSDP 126	34°27'	20°07'	2858
RC9-190	DSDP 132	38°39'	19°14'	1712
KSO9	DSDP 372	35°09'	20°09'	2800
TR 171-27	DSDP 374	33°50'	25°59'	2680
TR 172-22	DSDP 375	35°19'	29°01'	3150
Cobblestone (4) 6	DSDP 376	36°15'	17°42'	3466
Cobblestone (4) 42	DSDP 378	36°16'	17°43'	3592
Cobblestone (4) 44		36°16'	17°43'	3489
Cobblestone (4) 45		36°16'	17°43'	3466
Cobblestone (3) 29		35°50'	20°50'	2866
Cobblestone (3) 30		35°50'	20°51'	2885
Cobblestone (3) 32		35°51'	20°53'	3243
Cobblestone (3) 37		35°53'	20°47'	3038
		38°52'	04°60'	2726
		34°37'	20°26'	2782
		35°10'	21°26'	3730
		40°16'	11°26'	2835
		40°02'	04°02'	2699
		35°51'	18°12'	4078
		34°46'	31°46'	1900
		34°52'	31°48'	2101
		35°56'	25°07'	1835

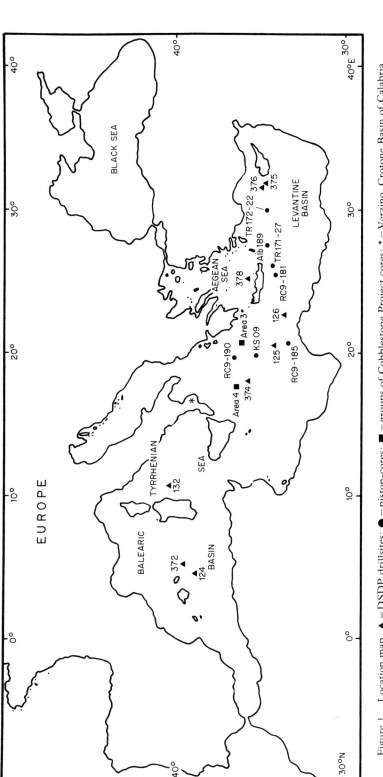

Figure 1. Location map. ▲ = DSDP drillsites; ● = piston-cores; ■ = groups of Cobblestone Project cores; * = Verzino, Crotone Basin of Calabria.

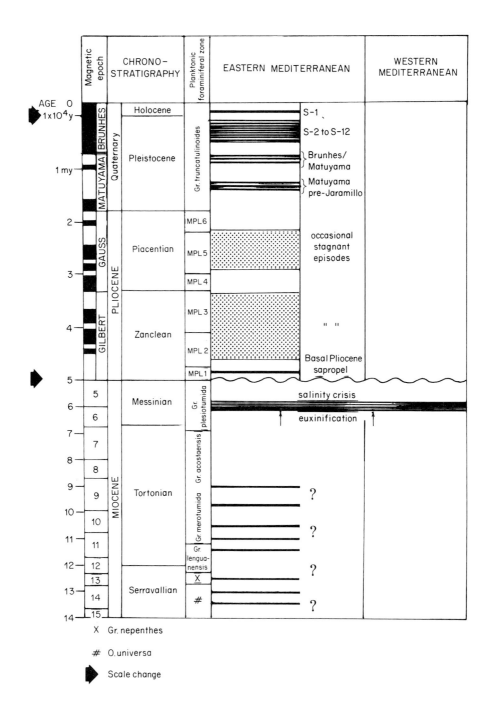

Figure 2.   Stratigraphic scheme of the Mediterranean Neogene, compiled from various sources, showing the position of stagnant episodes in the eastern and western Mediterranean basins, versus a calibrated biostratigraphy and chronostratigraphy.

Table 2.   Organic carbon percentages, pyrolisis assays, kerogen composition and pyrochromatographic analysis of 16 sapropel samples analyzed from eastern Mediterranean drillsites (DSDP Sites 374, 375, 376 and 378)

Site	Core	Section	Depth from top of section (cm)	Stratigraphy	Organic carbon (%)	Hydrocarbons produced by pyrolysis	HK/CO* (%)	Kerogen				Pyrochromatographic analysis			
								AOM†	MPh‡	CHF"	CWF■	Natural sample		Washed sample	
												Gas	Cond.	Gas	Cond.
374	3	1	140	Pleistocene	8.75	27 311	31.2	90	5	5	—	1084	77	1064	65
375	4	4	43	Tortonian	6.08	17 683	29.1	90	5	5	—	374	26	231	26
	6	3	47	Tortonian	1.54	686	4.5	85	5	5	5	54	—	26	—
	9	3	51	Serravallian	2.82	12 869	45.6	90	—	5	5	124	2	121	5
	9	3	133	Serravallian	0.38	228	6.0	70	5	10	15	28	—	28	—
376	1	4	79	Pleistocene	5.32	4669	8.8	80	10	5	5	773	32	773	32
	2	2	131	Pleistocene	6.09	7455	12.2	95	—	5	—	925	38	925	38
	2	2	134	Pleistocene	5.14	6739	13.1	90	—	10	—	964	38	964	38
	5	2	113	Pleistocene	4.02	1645	4.1	95	—	5	—	120	6	120	6
	5	4	103	Late Pliocene	2.80	11 874	42.4	95	—	5	—	332	32	262	32
	6	4	68	Early Pliocene	3.82	13 445	35.2	75	5	10	10	283	12	283	12
378	1	1	131	Pleistocene	6.59	18 410	27.9	95	—	5	—	433	26	433	26
	1	2	61	Pleistocene	5.66	17 094	30.2	95	—	5	—	581	34	531	34
	6	2	16	Late Pliocene	2.95	6408	22.5	95	5	—	—	157	80	157	80
	6	3	94	Late Pliocene	5.93	14 991	25.3	95	—	5	—	346	21	346	21
	8	2	36	Late Pliocene	3.21	13 147	41.0	95	—	5	—	289	22	289	22

* HK/CO = Hydrocarbon/Organic Carbon. † AOM = Amorphous Organic Matter. ‡ MPh = Marine Phytoplankton. " CHF = Continental Herbaceous Fitoclasts. ■ CWF = Continental Wooden Fitoclasts.

phytoclasts. One of the samples analyzed is so low in organic carbon that it cannot be considered a true sapropel: this is the sample in which continental phytoclasts are more abundant, an observation which matches those concerning the youngest sapropel investigated, as discussed later on. The marine origin of organic matter is in good agreement with the pelagic nature of the sediments in which the sapropel layers are included.

The oxygen-deficient depositional environment limited to the eastern Mediterranean is considered to be the response to the interruption of the communication with the Indian Ocean in mid-Miocene time, a result of the counter-clockwise rotation of the African plate.

Oxygen depletion increased drastically prior to the Messinian salinity crisis, extending to the western basins. Progressive reduction of the interconnection between the Atlantic and the Mediterranean close to the end of the Miocene, induced by continental collision near the Iberian portal, resulted in euxinification of the Mediterranean and in its eventual desiccation when it was completely cut off (Hsü *et al.*, 1973; Cita *et al.*, 1978; Hsü *et al.*, 1978; Cita, 1979). Other authors, including van Couvering *et al.* (1976) prefer a lowering of sea-level, induced by glaciation in Antarctica, as a causative mechanism for the progressive isolation of the Mediterranean; in any case, the result is a reduction of the deep thermo-haline circulation, followed by restriction of the basin, and its euxinification.

Euxinic facies sediments, finely laminated, non-bioturbated, often diatomitic are well-known from the land record where they directly underlie the Messinian evaporites (Tripoli Formation of Sicily, and correlative units). They have been investigated in detail from Sicily (Gersonde, 1978, 1980; Bennet, 1979; McKenzie *et al.*, 1979), from Algeria (Anderson, 1933; Gersonde, 1980), from Cyprus and Morocco (Bizon *et al.*, 1979). Diatoms are recorded in very high numbers (up to 90 million per gram of dry sediment), indicative of high productivity, in rapidly-deposited layers which alternate with more slowly deposited carbonate-rich intervals. Climate modulation of the cyclically repetitive alternation of diatomites and marls is accepted by most authors (McKenzie *et al.*, 1979; Meulenkamp *et al.*, 1979).

The sedimentologic and paleoecologic characters of these laminites suggest a sedimentation in a layered water mass, with reducing conditions at the bottom and high surficial productivity (Rouchy, 1980). The cyclic repetition involves periodic oxygenation of the bottom water resulting from the re-establishment of deep circulation.

These euxinic, pre-evaporitic facies were not cored at any of the drillsites of either Leg 13 or 42A of the Glomar Challenger in the Mediterranean: indeed, the only holes which penetrated the Mediterranean Evaporite (372 in the Balearic Basin and 375 in the Levantine Basin) are on basin margins where the formation pinches out, and an erosional gap prevents the observation of the euxinic, restricted stage.

The Tripoli Formation may occasionally be very rich in organic carbon, and also in hydrocarbons. Figure 3 shows an outcrop of "black" Tripoli near Verzino, in the Crotone Basin of Calabria (see Martina *et al.*, 1979). The approximately 5-metres thick section measured and sampled consists of an irregular alternation of organic carbon-rich (black) and organic carbon-poor (white) diatomites and marls, which were investigated analytically (Table 3). The 80-cm thick bed, jet black in color, near the base of the outcrop yielded 1.97 g of hydrocarbons per 100 g of sample. Extracted hydrocarbons are highly viscous, and have a bulk density at 15°C of 1.022 corresponding to 7.0°API. Composition of kerogen from this sample is dominated by amorphous organic matter of probable marine origin (70%). Organic matter is entirely continental in a piece of coal (sample 4855) embedded in a chaotic interval (see insert of Figure 3) which also contains blocks of a sulphur-bearing limestone

Table 3. Analytical data on organic carbon-rich and organic carbon-poor levels from the Tripoli Formation at Verzino, Calabria. Sample location is shown in Figure 3

Sample	Lithology	Distance from baseline (cm)	Grain size (%) 63–93 μ	93–200 μ	>200 μ	CaCO₃ (%)	Kerogen (%) AOM*	CHF†	CWF‡	Remarks
4858	cm-thick limestone	+405	not measured			6	60	10	30	Poorly preserved diatoms
4857	black laminated diatomite	+362	0.05	0.00	0.00					
4856	pale brown marl	+280	0.38	0.67	0.03	73				Allochthonous
4855	limestone, sulphur-bearing	+225	not measured							Allochthonous
4855 bis	lens of jet-black coal	+175	not measured				—	—	100	Allochthonous
4854	marl	+90	0.11	0.43	0.43	74.5				Marine diatoms indicative of restricted environment
4853	black diatomite	+40	0.35	0.00	0.00	16.5	70	5	25	Rich in hydrocarbons poorly preserved diatoms
4852	white calcareous marl	−55	0.33	1.97	0.24	80				
4851	sandy marl	−65	0.71	0.81	0.04	56.5				Poorly preserved diatoms

* AOM = Amorphous Organic Matter. † CHF = Continental Herbaceous Fitoclasts. ‡ CWF = Continental Wooden Fitoclasts.

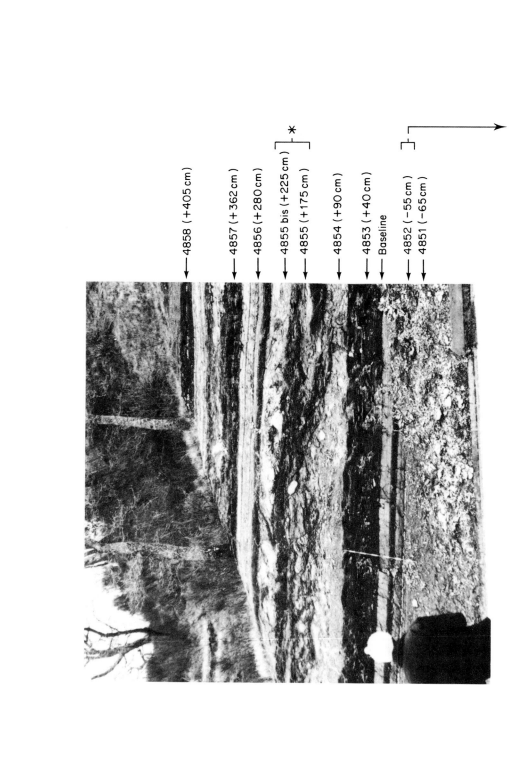

4858 (+405 cm)

4857 (+362 cm)

4856 (+280 cm)

4855 bis (+225 cm)

4855 (+175 cm)

*

4854 (+90 cm)

4853 (+40 cm)

Baseline

4852 (−55 cm)

4851 (−65 cm)

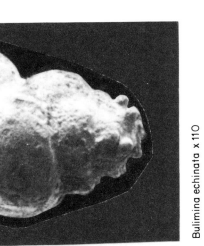

Bulimina echinata x 110

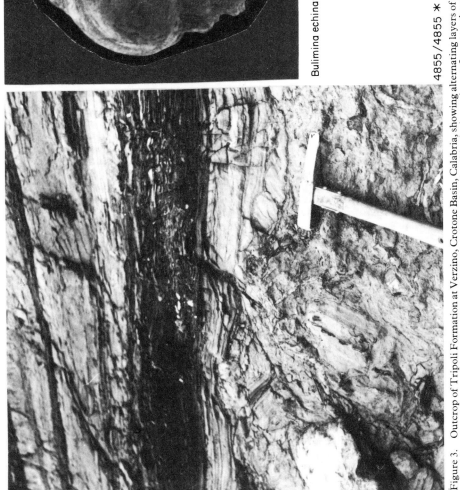

4855/4855 ✱

Figure 3.   Outcrop of Tripoli Formation at Verzino, Crotone Basin, Calabria, showing alternating layers of diatomites, sapropels and marls with a chaotic interval (samples 4855 and 4855 bis) depicted in detail in the lower insert. Insert to the right shows *Bulimina echinata*, a benthic foraminifer highly tolerant to stress conditions, which is abundant in sample 4852, beneath the lowermost sapropelic layer.

(sample 4855 bis) up to 40 cm in size. Gersonde (in Martina *et al.*, 1979) recorded evidence of some influx of river water in the diatom assemblage in sample 4854, as documented by the fresh water species *Melosira granulata* (planktonic) and *Cocconeis placentula* (benthonic). *Bulimina echinata* (insert in Figure 3), a benthonic foraminiferal species which heralds the onset of the salinity crisis (d'Onofrio and Iaccarino, 1978) is recorded in abundance in sample 4852 from below the hydrocarbon-rich thick black layer.

The example illustrated on the Tripoli from Verzino shows the great variability of situations that can be recorded in an apparently regular succession.

Sapropelic episodes also occurred repeatedly during the Messinian salinity crisis. We can use as an example Site 374, located just in the center of the Messina abyssal Plain. With a water depth of some 4100 m, it is the deepest abyssal plain of the entire Mediterranean.

Diatomites and mudstone intercalated in between the cyclically repeated evaporites are black, and sometimes organic-rich (Kidd *et al.*, 1978; Sigl *et al.*, 1978; Deroo *et al.*, 1978). The environment was entirely anoxic, even though not persistently organic-rich, due to the scarcity of biogenous material in the depressed water body.

Black, organic-rich intra-evaporitic diatomites are also recorded in the western Mediterranean, as at Site 124 from the Balearic Basin (Hajos, 1973; Schrader and Gersonde, 1978).

3. Climatic versus eustatic significance of Pliocene and Early Quaternary sapropels

After the termination of the Messinian salinity crisis, development of organic-rich facies is basically restricted to the eastern Mediterranean once again.

Early Pliocene sapropels were recorded in the Messina abyssal Plain (Site 374), in the Levantine Basin west of Cyprus (Site 376) and in the Cretan Basin north of Crete (Site 378). *There is no evidence that the process was climatically modulated.* Indeed, the Early Pliocene is a steady, relatively warm interval, predating the Late Pliocene climatic deterioration (Ciaranfi and Cita, 1973; Shackleton and Cita, 1979; Cita and Ryan, 1979; Thunell, 1979*a*).

Why did the eastern Mediterranean become stagnant immediately after the deep-sea Pliocene transgression? In the scenario of the deep basin desiccation model, this is considered the response to a structural setting, as strong evidence supporting the existence of a high-standing barrier (see Cita and Ryan, 1973, their Figure 3) separating the western Mediterranean, where the earliest Pliocene oozes, as recovered in the Tyrrhenian Basin (Site 132) indicate a well-oxygenated environment, and the eastern Mediterranean (Sites 374 and 376) where these sediments indicate anoxia at the bottom. The oldest sapropel recorded in the eastern Mediterranean belongs to the *Sphaeroidinellopsis* acme-zone of the basal Pliocene, and lies some 15 cm above the sharp contact with Late Messinian brackish sediments. Analysis on kerogen from this oldest Pliocene sapropel (376-6-4, 68 cm in Table 2) indicate a dominance of amorphous organic matter (75%); the marine origin is supported by the very high HK/CO ratio. Content in organic carbon is 3.82%.

Additional evidence for a shallow sill between the western and the eastern Mediterranean basins is provided by clay mineralogy (Chamley *et al.*, 1978). The proportion of chlorite versus other clay minerals increases in Upper Messinian sediments at the eastern Mediterranean drillsites, but not in the western ones. This might indicate a specific geochemical environment, propitious to some magnesium-rich mineralogical growth. The latter assumption is supported by the occurrence of

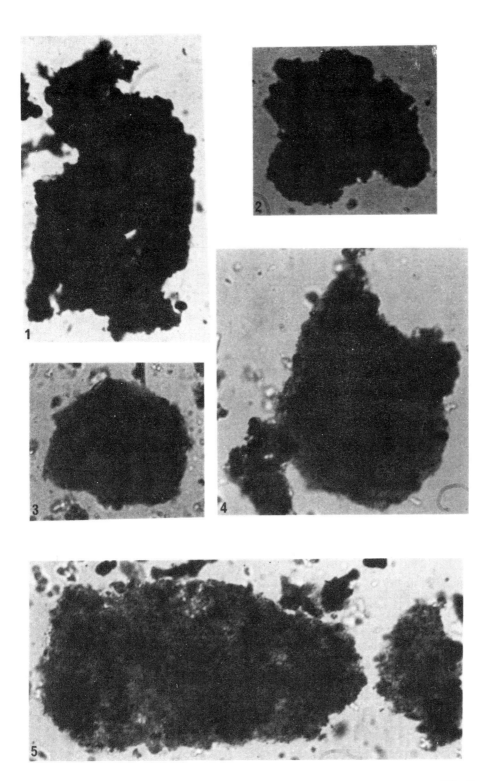

Plate 1. (1) Sapropel S 1—Core 37/1 cm 36 (×450); (2) Sapropel S 1—Core 44/3 cm 111 (×450); (3) Sapropel S 1—Core 32/5 cm 143 (×450); (4) Sapropel S 1—Core 32/5 cm 6 (×450); (5) Sapropel S 1—Core 32/5 cm 88 (×350).

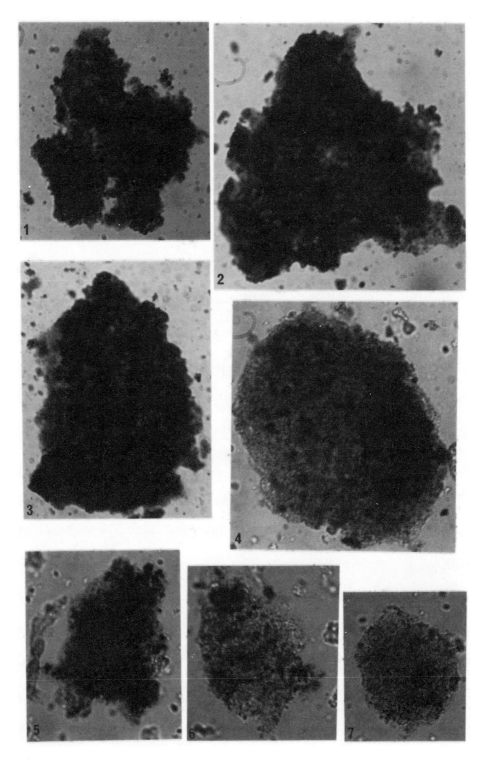

Plate 2. (1) Sapropel S 5—Core 30/2 cm 50 (×350); (2) Sapropel S 5—Core 29/2 cm 47 (×450); (3) Sapropel S 5—Core 45/7 cm 80 (×450); (4) Verzino—Sample no. 4853 (×450); (5) Verzino—Sample no. 4857 (×450); (6) Verzino—Sample no. 4857 (×450); (7) Verzino—Sample no. 4857 (×450).

idiomorphic dolomitic crystals both in the eastern Mediterranean DSDP boreholes and in Late Messinian sediments from the Apennine foredeep (Casati *et al.*, 1978; Colalongo *et al.*, 1978) and from the Ionian Islands (Vismara Schilling *et al.*, 1978).

There is some evidence that Early Pliocene sapropels are more widespread in (or limited to) basinal settings, whereas they seem to be missing on structurally elevated areas (Kidd *et al.*, 1978). This suggests that the boundary layer separating surficial waters with dissolved oxygen from deep waters with dissolved hydrogen sulphide was lower in the Early Pliocene than the approximately -700 m calculated for the Holocene sapropel (van Straaten, 1972; McCoy, 1974). The record is not exhaustive, however, since both drill sites with continuous coring in the eastern Mediterranean present some shortcomings: Site 125 (Mediterranean Ridge crest) because core recovery did not exceed 50%; Site 376 (Antalya Basin) because of the irregular succession recovered.

The earliest part of the Late Pliocene (foraminiferal Zone M Pl 4) whose biostratigraphic boundaries are calibrated via paleomagnetic stratigraphy at 3.3 and 3.0 m.y. respectively (Cita, 1975) is characterized by a conspicuous absence of sapropels. It is the time during the last 5 m.y. when the eastern Mediterranean was best ventilated. This was considered an indication of intensified circulation induced by the climatic deterioration leading to the development of continental ice sheets in the northern hemisphere, and consequent glacio-eustatic lowering of sea-level (Kidd *et al.*, 1978; Keigwin and Thunell, 1979; Thunell, 1979*a*). This explanation seems no more appropriate (Mörner, 1980), since the first expansion of grounded ice sheets occurred later. The re-activated deep-sea circulation might be attributed to the closure of the Panama isthmus, which disconnected the Pacific from the Atlantic, enhancing north–south gradients and circulation in the latter (Mörner, pers. commun., 1980).

After this short period of enhanced oxygenation related to increased thermo-haline circulation at depth, the eastern Mediterranean once again became stagnant in the Late Pliocene. Late Pliocene sapropels were cored at DSDP Sites 125, 374, 376 and 378. The highest value of organic carbon (over 16% by weight, corresponding to 30% organic matter) was recorded from the late Pliocene foraminiferal zone M Pl 5, from a sapropel recognized in the Messina abyssal Plain drill site (Sigl *et al.*, 1978).

Organic carbon contents from Late Pliocene sapropel samples analyzed in this study (see Table 2) range from 2.80–5.93%. In all these samples the HK/CO ratio is very high, suggesting a marine origin, which is also supported by the strong dominance of amorphous organic matter (AOM) in kerogen (95%). The same dominance of AOM is also recorded in sapropels from the Cretan Basin (Site 378) which differ from those recovered in other eastern Mediterranean basins in several aspects: (a) they are much more frequent in the column, (b) their color is greenish or brownish instead of black, (c) they contain sparse benthic foraminifers and are burrowed throughout, thus indicating that there was no complete anoxia at the bottom, and (d) some of them contain diatoms. One could suspect that organic matter had a continental origin, also considering the position of the Cretan basin, surrounded by dozens of vegetated islands, but the petrographic analysis of kerogen disproves this hypothesis. The hydrogen/carbon ratio is very high in all the samples analyzed, thus supporting the marine origin of organic matter. The frequency of sapropel layers could suggest astronomic-type cycles, but the stratigraphic control is inadequate (no continuous coring).

Sapropels are also recorded from the Early Pleistocene ("Matuyama pre-Jaramillo sapropels" of Cita and Ryan, 1973 and Cita *et al.*, 1973), and in the lower part of the *Pseudoemiliania lacunosa* nannofossil zone as well, close to the Brunhes/Matuyama boundary ("Brunhes/Matuyama sapropels"). As for the Late Pliocene sapropels, they were first discovered during Leg 13 of the Deep Sea Drilling Project, and have subsequently been recorded in a few cores of the Cobblestone Project* which penetrated deep into the Quaternary section along the slopes of an area of the western Mediterranean Ridge characterized by strong vertical relief (Blechschmidt *et al.*, 1982).

In conclusion, Early Pliocene sapropels are interpreted as related to the post-Messinian transgression and do not seem to be controlled by climate, whereas the Late Pliocene and Early Pleistocene ones seem to be climatically modulated. A close comparison with younger sapropels, however, is hampered by the lack of information on eventual changes of composition of the planktonic assemblages across sapropel layers, and on the isotopic composition of oxygen in foraminiferal tests.

4. Stagnant episodes of the ice ages

Late Pleistocene and Holocene sapropels are the best known. They have been identified and correlated in dozens of piston cores (Olausson, 1961; Ryan, 1972; McCoy, 1974; Sigl and Müller, 1975; Maldonado and Stanley, 1976; etc.).

Sapropels—sedimentary expression of stagnation—and tephra layers—sedimentary expression of explosive volcanic activity—represent isochronous lithologies and permit a high resolution stratigraphy to be applied to the eastern Mediterranean deep-sea record, with precise bed to bed correlation (Ryan, 1972; Ryan and Cita, 1977; Keller *et al.*, 1978).

Micropaleontologic and isotopic investigation on selected piston cores carried out in the last 25 years (Emiliani, 1955; Parker, 1958; Ryan, 1969; Luz and Bernstein, 1977; Cita *et al.*, 1977; Thunell *et al.*, 1977; Thunell, 1979*b* etc.) allowed calibration of the local record by correlation with that of the open ocean.

The most complete core ever recorded is RC9-181 from the Mediterranean Ridge south of Crete (Ryan, 1972; Vergnaud-Grazzini *et al.*, 1977). It contains twelve sapropels encompassing climatic stages Z through U (S-1 to S-11) (Figure 4). Most sapropels are located near negative peaks of the isotopic curve of oxygen, or along warming trends of the curve. Some sapropels occur immediately after well-documented transgressions, as exemplified by sapropels S-1 and S-5.

4.1. Stagnant episode related to the Tyrrhenian (~125 000 years BP) transgression

Sapropel S-5, with an interpolated age of approximately 125 000 years, corresponds to the Tyrrhenian transgression (Riss-Würm alpine interglacial; Eemian interglacial of northern Europe). It coincides with isotopic substage 5e. This sapropel records the warmest conditions ever experienced by the eastern Mediterranean in all the deep-sea cores we had the opportunity to investigate, as indicated by (a) the isotopic signal in Core RC9-181 (Vergnaud-Grazzini *et al.*, 1977), (b) the isotopic and faunal signals in Core KS 09 (Cita *et al.*, 1977) and (c) the faunal signal in cores 6, 29 and 45 of the Cobblestone Project (Cita *et al.*, 1982).

---

* This name is given to an international project whose at-sea operations were accomplished in 1978. The R/V *Melville* surveyed four discrete areas of the Mediterranean Ridge and Calabrian Ridge, each approximately 100 km² wide, characterized by a highly irregular bottom topography, using the deep-tow instrumentation of the Marine Physical Laboratory of Scripps Institution of Oceanography. Operations included detailed bathymetric survey, 4 kHz seismic reflection profiling, side scan sonar and bottom photography. Two of the areas surveyed with the deep-tow system in the Ionian Basin were visited one month later by R/V *Eastward* for a transponder-navigated coring program. Forty piston cores precisely positioned with reference to the highly irregular bottom topography were recovered, some of which are discussed here.

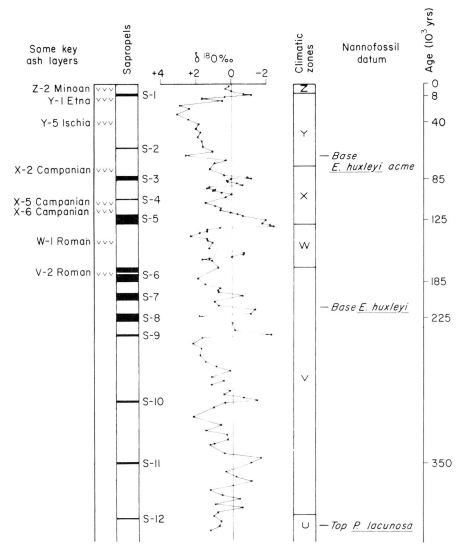

Figure 4. Columnar log and isotopic ratio of O (measured on *Globigerinoides ruber*) in Core RC9-181 (after Vergnaud-Grazzini *et al.*, 1977). Some key ash layers, climatic faunal zones, and nannofossil datum planes (after Blechschmidt *et al.*, 1982) are also shown.

Sapropel S-5 in Cobblestone Core 30 has the highest organic carbon percentage measured, up to 6.33% (Table 4). Optical analysis of kerogen showed large flakes of amorphous organic matter (Plate 2, Figure 1). An even higher value of organic carbon (6.63%) was measured in S-5 from Cobblestone Core 29 (Cita *et al.*, 1982). Figure 5 illustrates sapropel S-5 in Cobblestone Cores 29 and 30: notice the sharp contacts and the total absence of burrowing. The color of the normal, well-oxygenated sediments above and below the sapropel matches perfectly in these two cores, raised from two different plateaus of the Mediterranean Ridge.

In a pilot-study on destructive effects of oxygen starvation on benthic life and the rate of repopulation (Cita and Podenzani, 1980), we found a sudden disappearance of all benthic foraminifers at the base of sapropel S-5 in Core RC9-185. Faunal density and faunal diversity went to zero, and we found no evidence whatsoever of

Table 4. Organic carbon percentages and kerogen composition in 18 samples analyzed from Cobblestone cores 32 (sapropel S-1), 44 (sapropel S-1), 37 (S-1), 30 (S-5), 29 (S-6) and 45 (S-6)

Core	Distance from top of core (cm)	Age	Lithology	Organic carbon (%)	Kerogen (%)			
					AOM	MPh	CHF	CWF
32	587	Holocene	sapropel S-1	1.55	70	0	15	15
32	598	Holocene	sapropel S-1	1.15	20	—	60	20
32	611	Holocene	sapropel S-1	1.43	70	—	20	10
32	631	Holocene	sapropel S-1	1.29	60	5	10	25
32	651	Holocene	sapropel S-1	1.46	60	—	15	25
32	653	Holocene	sapropel S-1	1.46	60	—	25	15
32	673	Holocene	sapropel S-1	1.15	50	—	20	30
32	693	Holocene	sapropel S-1	2.13	70	—	15	15
32	710	Holocene	sapropel S-1	2.26	80	—	—	20
44	256	Holocene	marl	0.28	5	5	20	70
44	261	Holocene	saprople S-1	2.26	70	—	10	20
44	266	Holocene	sapropel S-1	2.36	70	—	5	25
44	269	Holocene	sapropel S-1	2.40	70	—	5	25
44	275	Holocene	marl	0.37	20	—	40	40
37	36	Holocene	sapropel S-1	2.95	70	—	10	20
30	200	Pleistocene	sapropel S-5	6.33	80	—	10	10
29	197	Pleistocene	sapropel S-6	3.21	70	—	10	20
45	980	Pleistocene	sapropel S-6	4.43	70	—	10	20

Figure 5.   Sapropel S-5 in two piston-cores raised from two discrete plateaus in Cobblestone Area 3, western Mediterranean Ridge.

progressive adaptation of the fauna to environmental stress. The rate of repopulation was very slow; for at least one thousand years after the termination of the stagnant episode, re-colonization of the bottom had not started yet.

This finding is consistent with the isotopic data, which indicate that the climatic reversal coincident with the stage 6/stage 5 boundary (Termination II) was very rapid. In Core RC9-181 it records the largest temperature change identified in the eastern Mediterranean. According the Vergnaud-Grazzini *et al.* (1977) the isotopic change, corrected for glacial effect and for evaporation effect, corresponds to a temperature change of 8°C, versus the 4°C recorded at the stage 2/stage 1 boundary (Termination I) in the same core.

4.2. Stagnant episode related to the Flandrian (~ 9000 years BP) transgression

Sapropel S-1, with an interpolated age of 8000 years BP and dated radiometrically at approximately 9000 to 7400 years BP (Olsson, 1959; Olson and Broecker, 1959; Hieke *et al.*, 1973; Stanley and Maldonado, 1977, etc.), follows the post-glacial, Flandrian transgression.

Sapropel S-1, correlative to isotopic stage 1, is isotopically light in Core RC9-181 (Figure 4). It is faunistically warm in Cobblestone Cores 6 and 45 (Cita *et al.*, 1982).

Our study on benthic foraminiferal assemblages on Core RC9-190 (Cita and Podenzani, 1980) showed good evidence of adaptation of the bottom-living fauna to progressively deteriorating environmental conditions, with faunal density and faunal diversity decreasing as the sapropel is approached, and with the composition of the foraminiferal fauna reflecting a progressive oxygen deficiency at the sediment/water interface; the genus *Bolivina*, the foraminiferal genus most tolerant to low oxygen levels, takes the place of the previously dominating genus *Cassidulina* (Figure 6). The re-colonizing foraminiferal fauna is entirely different from that which populated the bottom prior to the post-glacial stagnant episode. Indeed, our pilot study, as well as further studies now in progress, shows that the present day benthic fauna of the eastern Mediterranean has not yet recovered entirely from the last stagnation.

Content of organic carbon in sapropel S-1, measured in several piston cores of the Cobblestone Project ranges from 2 to 3%.

Thickness of sapropel S-1 ranges from 10 to 20 cm, less in low-sedimentation-rate cores. The record thickness of 155 cm was measured in Cobblestone Core 32, from the center of Aphrodite Crater, a depression with a vertical relief of some 250 m. Only the lower part of the sapropel (dense, dark and non-laminated) is certainly the result of *in situ* biogenic deposition in anoxic conditions. The greater portion of the sapropel seems to be redeposited or diluted, and shows laminations (Figure 7). We sampled this sapropel at regularly-spaced intervals, analyzed its kerogen composition as well as mineralogical composition (by x-ray diffractometry), and studied with the same methodology an interval encompassing sapropel S-1 in Cobblestone Core 44, from the flank of a plateau in the southern Calabrian Ridge. The thickness of S-1 in Core 44 is one tenth that in Core 32. Figures 7 and 8 document the composition of kerogen and the x-ray mineralogy (see also Table 4 for the organic geochemistry).

Organic carbon of S-1 are comparable in Core 44 and in the basal part of Core 32, whereas they decrease to less than 2% in the upper part of the latter core, interpreted as resedimented. Figure 7 shows that the percentage of continental phytoclasts (both herbaceous and ligneous) is inversely correlated with organic carbon content. In normal sediment (Core 44) where organic carbon is less than 0.3%, continental phytoclasts dominate the kerogen, whereas they are strongly subordinate to

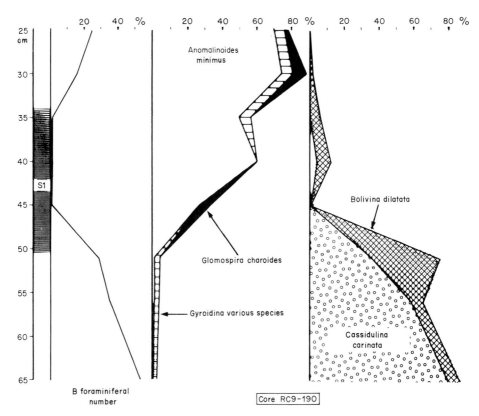

Figure 6.   Quantitative changes recorded in populations of benthic foraminifers in core RC9-190 across
        sapropel S-1 (after Cita & Podenzani, 1980, modified). B foraminiferal number is the number of
        benthic foraminiferal specimens per gram of dry sediment, counted in the sediment fraction greater
        than 63 microns.

amorphous organic matter in the sapropel. A peak of herbaceous continental
phytoclasts near the top of S-1 in Core 32 coincides with a local minimum of organic
carbon, and with a local maximum of chlorite, suggesting a stronger terrigenous
influx diluting the sapropel. No marked differences in the mineral composition are
noticed across sapropel S-1. The differences between Core 44 (southern Calabrian
Ridge) and 32 (western Mediterranean Ridge) are accounted to regional factors, the
higher concentration of carbonates recorded in the Mediterranean Ridge core
compares well with data obtained from other Cobblestone cores (Cita et al., 1982), as
well as the lesser percentages of smectite and kaolinite.

    Figure 9 illustrates sapropel S-1 in Cobblestone Core 44: the darker hues noticed
near the base of the sapropel correspond to slightly higher organic carbon content
(see Figure 7).

    The color of organic matter is much darker in Core 44 than in Core 32 (see Plate 1,
Figures 2–5). Color of organic matter is generally considered evidence of maturation,
with darker hues indicative of stronger heating, thus of older age and /or more
extended burial. Our results are not easily explained in the light of these widely
accepted concepts; in fact, the age is exactly the same, and S-1 is shallower in Core 44
(260 cm from core top) than in Core 32 (580 cm from core top). A differential heating
can hardly be an explanation for the color difference, since (a) the crust on which
both areas lie is old, and continental, and (b) heat flow is very low in the Ionian Basin
(Erickson & van Herzen, 1978).

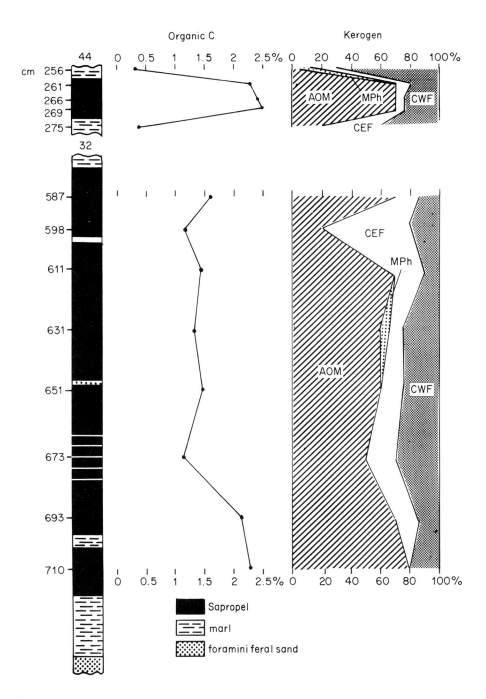

Figure 7. Organic carbon content and kerogen composition recorded in sapropel S-1 and adjoining strata in core Cobblestone 44 (top), southern Calabrian Ridge, and 32 (bottom), western Mediterranean Ridge.

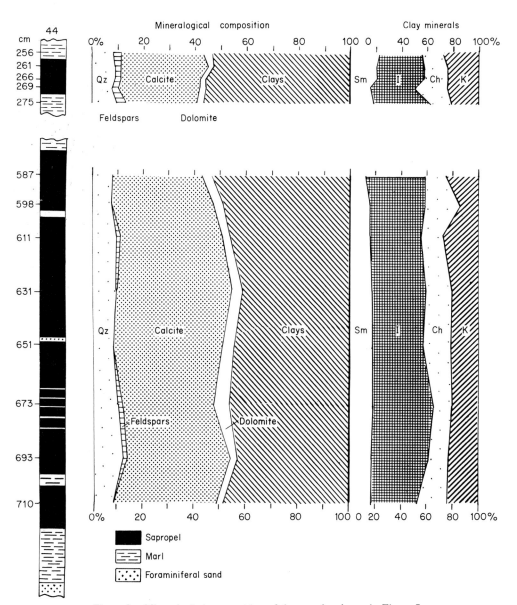

Figure 8.   Mineralogical composition of the samples shown in Figure 7.

Although we are unable to answer the last two points raised i.e., (1) where did the organic matter come from and (2) what other process can control its apparent maturation, we can conclude that sapropels such as those discussed so far, S-5 or S-1 fall into the category of those related to transgressions caused by eustasy in times of rapid deglaciation, which led to oxygen deficiency. In a semi-enclosed basin such as the eastern Mediterranean, separated from the open ocean by two shallow thresholds (the Gibraltar Straits and the Sicily Channel), basinwide stagnation was inescapable.

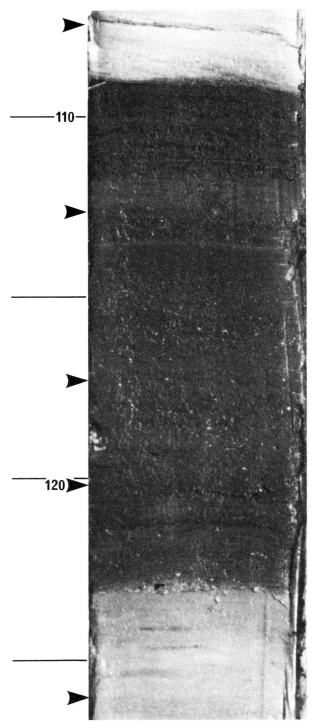

Figure 9.   Close up of sapropel S-1 in core Cobblestone 44. Notice uninterrupted nature of the sapropel and absence of burrowing.

*4.3. Stagnant episodes related to density stratification
essentially due to low salinity surficial waters*

We now consider another category of sapropels: those which are not on warm peaks, or along warming trends of isotopic or faunal curves. Some of the stage 7 sapropels, such as S-6 or S-8 are good examples.

In Albatross Core 189, investigated over 25 years ago by Emiliani (1955) and Parker (1958), positive values of the isotopic curve, and cold surficial conditions indicated by populations of planktonic foraminifers coincide with peak abundances of *Globigerina eggeri* (=*Neogloboquadrina dutertrei*), a species which is known to tolerate low salinities (Ruddiman, 1971). This species is usually present in very low percentages in the eastern Mediterranean deep-sea record, but its abundance increases drastically in several sapropels (not in all the sapropels, however) and reaches peaks in excess of 50% in S-6 and S-8. The species can thus be used as a salinity indicator (Vergnaud-Grazzini *et al.*, 1977; Thunell *et al.*, 1977: see also further discussion).

In the plateau cores of the Cobblestone Project, the salinity faunal index (see Figure 10) permits a good correlation at the level of sapropel S-6, which is faunistically cold.

Furthermore, in a study recently accomplished on the distribution of benthic foraminifers in Core KS 09 (Parisi and Cita, in press) drastic changes in density, diversity and composition of the bottom-living forms were recorded (Figure 11). Also with this approach, a fundamental difference was noticed between sapropels S-6 and S-8 and all the remaining ones occurring in that core: the species *Bulimina exilis, Cassidulinoides tenuis, Ellipsopolymorphina* sp., *Fursenkoina complanata* occur in some abundance in discrete levels above these two sapropels (recolonizing fauna), then disappear entirely.

If oxygen deficiency resulting from glacio-eustatic sea-level rises (transgressions) is assumed, how can the "cold" sapropels be explained?

5. Discussion

Since the first discovery of sapropels in the late Quaternary deep-sea record of the eastern Mediterranean, and especially since the Swedish Deep-sea Expedition of 1947–1948, the nature and origin of organic carbon-rich sediments have generated much interest and debate. We do not review here the many papers dealing with the topic, but focus our discussion on basic points.

The Mediterranean is an example of anti-estuarine circulation (Tchernia and Lacombe, 1972). It is filled with water masses comparable to ocean surface water in terms of their high degree of $CaCO_3$ saturation and low concentration of silica and nutrients.

Salinity in the Mediterranean is higher than in the open ocean, up to 39.5% and is enriched in $\delta^{18}O$ by up to 1.5% (Vergnaud-Grazzini, 1973). Magnesium calcite is being precipitated (Milliman and Müller, 1973 and later workers). All these are considered to be effects of excess evaporation.

During times of deglaciation, all the water which entered the eastern Mediterranean (a) from the Atlantic, via the western Mediterranean, (b) from precipitation, (c) from river runoff, and (d) from glacial runoff (essentially via the Black Sea, since there were no continental ice sheets drained directly by the Mediterranean), was less dense than the local Mediterranean water, so that density stratification and consequent stagnation was unavoidable.

According to some authors, including Berger (1976), during times of increased

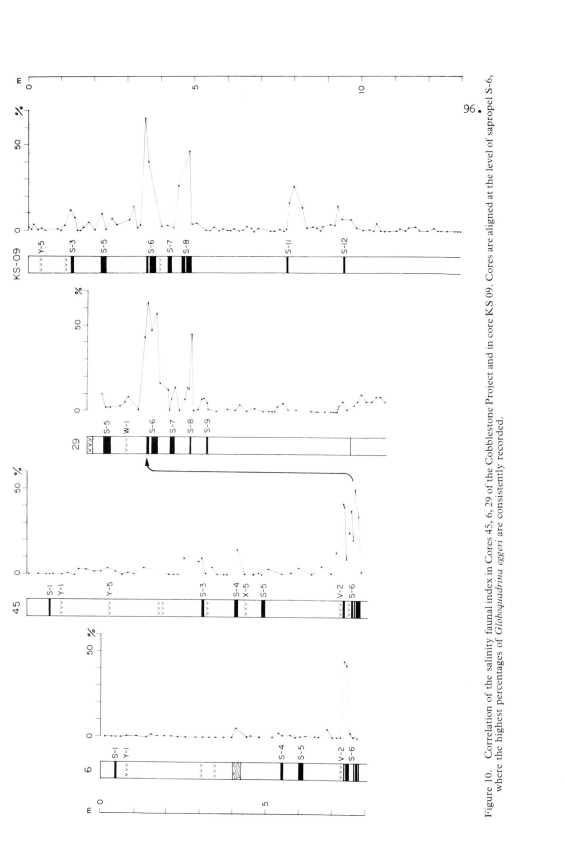

Figure 10. Correlation of the salinity faunal index in Cores 45, 6, 29 of the Cobblestone Project and in core KS 09. Cores are aligned at the level of sapropel S-6, where the highest percentages of *Globoquadrina eggeri* are consistently recorded.

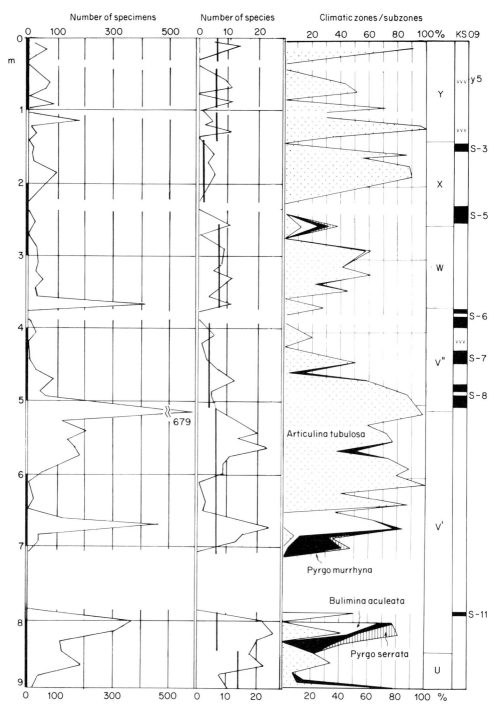

Figure 11. Changes in faunal density (left column), faunal diversity (second column from the left) and percentages of selected species of benthic foraminifers in the first 900 cm of core KS 09. The columnar log is shown to the right, along with the distinction of climatic zones (see also Figure 4). The subdivision of zone V is after Parisi & Cita (in press). The solid line in the column plotting the number of species shows the mean value calculated for each climatic zone and/or subzone.

precipitation and runoff, the deep circulation may have reversed, becoming estuarine; co-occurrence of diatoms and sapropelic mud would be a necessary consequence of the circulation reversal.

Perhaps the more controversial aspect of the various interpretations is the role of salinity in controlling anoxia, versus the role of temperature.

According to Arthur (1979) substantial fresh water runoff results in the formation of a brackish "lid" over an isolated basin. No attempt to quantify the process has been made so far. How thick should such a lid be in order to persist for a few thousand years without being destroyed by the strongest gales? Is the glacial runoff compatible with this minimum thickness?

The salinity effect has been stressed in several recent papers (Williams et al., 1978; Thunell, 1978; Williams and Thunell, 1979) with special reference to sapropels A ($\sim$9000 yrs BP$=$our S-1) and B ($\sim$80 000 yrs BP$=$our S-3). Salinity reductions would have been of the order of 2–3% and large increases of temperature are refuted.

On the other hand, Nesteroff (1973) suggested that stagnation may have been induced by a thermal stratification when surface waters warmed more rapidly than the local bottom water. Berger (1976) interprets *Neogloboquadrina dutertrei* as an indication of upwelling, not of low salinity, and considers "incidental" salinity changes during sapropel deposition.

Do we know any more now than it was known in the sixties, when Olausson (1965) followed by Ryan (1969) proposed the model of Black Sea–Mediterranean interactions in which the influx of fresh water from glacial melting was fundamental?

5.1. Magnitude of isotopic excursions

The excursions of the isotopic signal of O are much larger in the eastern Mediterranean than in the open ocean. The isotopic records obtained by Luz and Bernstein (1977), Cita et al. (1977), Thunell et al. (1977), Vergnaud-Grazzini et al. (1977) on several piston cores raised from the Ionian and Levantine basins confirmed the original observations by Emiliani (1955) in Core Albatross 189. This suggests that besides the global signal, we have a local Mediterranean signal superimposed on the curve, and that this signal is salinity-controlled.

5.2. Black Sea deep-sea record

The Black Sea was a fresh-water lake for the entire duration of the Pleistocene. Deep-sea drilling showed that only in Holocene time the Black Sea was invaded by Mediterranean waters (Ross, Neprochov et al., 1978). Consequently, we can argue about the amount of spill-over during high sea-level stands versus low sea-level stands, but consistently no return from the Mediterranean into the Black Sea.

5.3. Deep-sea record of the Marmara Sea

Based on a detailed study of four piston-cores from the Marmara Sea (Stanley and Blanpied, 1980), whose stratigraphic record extends back in time to 13 530 yrs BP (maximum radiocarbon age measured), we know that the lithofacies document fluctuating patterns of water mass exchange. Sediments are essentially non-carbonate ($<5\%$ $CaCO_3$) and rich in organic matter (11–19%); they accumulated very fast (sedimentation rates up to 70 cm per 1000 yrs). According to the above authors, conditions in the Marmara Sea were lacustrine prior to 12 000 yrs BP; partially anoxic from 12 000 to 9500 yrs BP; poorly ventilated from 9500 to 7000 yrs BP.

5.4. Occurrence of diatoms in Mediterranean sapropels

Diatoms are only recorded in sapropels from the Aegean Sea, including the Cretan
Basin DSDP Site 378, and in a definite area of the Levantine Basin, whereas they are
conspicuously absent in all the sapropels of all the piston-cores investigated so far
from the Ionian Basin (Kidd *et al.*, 1978; Schrader and Matherne, 1981; Cita *et al.*,
1982).

5.5. Characterization of low-salinity foraminiferal faunas

Planktonic foraminiferal faunas from discrete sapropels of different stratigraphic
position differ greatly both qualitatively and quantitatively (see Figure 10).
Observations by Parker (1958) on Albatross 189, further elaborated by Ryan (1969,
1972), were reproduced in several deep-sea cores from both the Ionian and the
Levantine Basins (Cita *et al.*, 1973; Cita *et al.*, 1977; Williams and Thunell, 1979; Cita
et al., 1982). Also the benthonic foraminiferal faunas pre-dating and post-dating
discrete sapropels are strongly differentiated (Parker, 1958; Cita and Podenzani,
1980; Parisi and Cita, in press).

5.6. Maximum glacial expansion versus last glacial expansion

The maximum glacial expansion is not the last one. This has been known for many
years (see for example Flint, 1957), but it has been more definite and clearer recently,
especially since a certain communication between marine stratigraphers and land
geomorphologists working on the Quaternary commenced (Bowen, 1978).

The last three points are fundamental for our discussion. Point 5.4. discards Berger's
(1976) hypothesis that Mediterranean sapropels in general are due to upwelling. In
particular, the concept that *G. eggeri* (= *N. dutertrei*) is an indication of upwelling has
to be abandoned. Peak abundances of the species are never accompanied by the
occurrence of diatoms, or other siliceous remains. Schrader and Matherne (1981)
investigated six eastern Mediterranean cores encompassing several well-identified
sapropels. These cores include RC9-181 with its twelve sapropels (see Figure 4) and
Cobblestone 30 and 37, with sapropels S-5 through S-8. Diatoms and silicoflagellates
were only observed in cores TR 171-27 and TR 172-22, within sapropel layers C
(= our S-4) and D (= our S-5). Diatoms were found to be common only in sapropel
D of a core from the Levantine Basin. The model proposed by Schrader and
Matherne (1981) considers increased primary productivity, induced by upwelling
and leading to increased oxygen consumption, as causative of anoxia at the bottom.
The applicability of this model, however, is very limited since, in most cores
analyzed, S-5 is diatom-free.
  The above demonstrates that the eastern Mediterranean became stagnant after the
Tyrrhenian transgression regardless of the existence of local upwelling conditions.
In other words, the upwelling mechanism controlled by local wind patterns in the
eastern corner of the Mediterranean is superimposed on a more general and regional
phenomenon which—unlike the former—is basin-wide (Ryan and Cita, 1977).
  A similar situation seems to apply to the Cretan Basin. Although the precise
stratigraphic position of diatom-bearing sapropels is unknown, since DSDP Site 378
was spot-cored in its upper part, we concur with Kidd *et al.* (1978) that diatom
blooms were induced by sudden upwelling of phosphorous-saturated deep waters
resulting from unusual wind patterns or other climatic circumstances. With 95%
amorphous organic matter (see Table 2), these Cretan Basin sapropels are certainly
marine in origin.

Point 5.5. discards the idea recently put forward by Thunell and Lohman (1979) of an "average sapropel planktonic fauna", based on a sophisticated statistical treatment of numerical data from twelve sapropels of unidentified stratigraphic position, and from 66 sediment samples. It supports the hypothesis that at least two discrete processes can lead to anoxia, one of which is related to transgressions induced by a rise in sea-level during times of deglaciation, and is recorded along warming trends of the climatic curves, whatever means we use to create these curves (isotopic ratio of oxygen, simple counting of planktonic foraminifers, factor analysis based on transfer functions of Mediterranean faunas). The second process is controlled by salinity changes probably much greater than the 2–3 per mil invoked by Williams and Thunell (1979).

Point 5.6. discards the hypothesis strongly stressed by Williams *et al.* (1978) and Williams and Thunell (1979) that low salinity stratification, induced by meltwater produced by disintegration of the Fennoscandian Ice Sheet, is primarily responsible for the last (Holocene) stagnation. Indeed, according to Grosswald (1980) the Eurasian Ice Sheet extended during the Late Weichselian glaciation (= Wurm alpine glaciation) from Iceland to the northern Taimyr Peninsula, over a distance of some 6000 km, reaching as far south as 52°N in western Europe. The ice sheet, the volume of which is estimated at 14×10^6 km³, impounded the north-flowing rivers and caused formation of ice-dammed lakes in west Siberia and the northern Russian Plain. The proglacial drainage system had a radial pattern that ended in the eastern Mediterranean via the Black Sea. After 13 500 yrs BP the system became marginal and discharged water into the Norwegian Sea. Therefore, when sapropel S-1 was being deposited 9000 to 8000 yrs BP, meltwater from the Eurasian Ice Sheet could not enter the Mediterranean, feeding a surficial low-salinity layer.

The case is different for the older sapropels such as S-6 or S-8, whose interpolated ages are approximately 185×10^3 yrs and 225×10^3 yrs respectively (see Figure 4). This is the time of the maximum glacial expansion, when the continental ice sheet extended as far south as 48°N in the Dnieper Lobe and less than 50°N in the Don Lobe (Flint, 1957; see also Figure 7 of Vergnaud-Grazzini *et al.*, 1977). It is the so-called "Dnieper" glaciation, whose acme has a numerical age of some 200×10^3 yrs (Nikiforova, pers. commun.). That the Don Lobe and the Dnieper Lobe belong to the same glaciation has been questioned (Velitchko, 1977). The former is now considered older (Oka or Mindel glaciation) than the latter (Dnieper or Riss glaciation). In any case, it is clear that discharge of meltwater from the retreating ice sheets was then in the eastern Mediterranean via Black Sea.

Other points to be stressed include the following.

5.7.   Depth of the oxygen-rich/oxygen-poor interface

McCoy (1974) isopached S-1 in the Ionian and Levantine Basins, and observed that the Holocene sapropel is never recorded in piston-cores from depths shallower than 700 m. Van Straaten (1972) arrived at a similar conclusion for the southern Adriatic. Sapropel S-1 is certainly not the most prominent sapropel detectable in piston cores, although it is that which is recorded more frequently. Other sapropels, such as for instance S-5, have consistently higher organic carbon content (see Table 4); moreover they lack bioturbations and show more drastic annihilation of bottom-living faunas (Cita and Podenzani, 1980). Consequently, we can assume that in S-5 times the depth of interface was shallower than in S-1 times, probably closer to the − 200 m of the present-day Black Sea, the prototype of euxinic conditions.

An observation supportive of this statement is the recorded absence of both mesopelagic species of planktonic foraminifers, for instance *Globorotalia truncatulinoides, G. inflata, Sphaeroidinella dehiscens*, and of bathypelagic ones (*Globorotalia scitula*). The only taxa recorded in sapropel S-5 live in the surface water layers (epipelagic) and this peculiarity of the faunas suggests that deeper waters were anoxic.

Sapropel S-5 can be reached by piston-coring only in low-sedimentation-rate pelagic settings, at locations that lie much deeper than the supposed depth of the interface. Consequently we cannot provide observational data which document the above assumption.

5.8.   Cyclicity of organic carbon-rich facies

Cyclic sedimentation in the Late Quaternary, including lower-than-normal salinities (eventually leading to sapropel deposition) and higher-than-normal salinities (high magnesium calcite levels), has been reported from the Nile cone (Maldonado and Stanley, 1976).

Of the twelve sapropels under discussion, those which seem to be the expression of a more or less regularly cyclic astronomically-modulated process, are (see Figure 4) (1) sapropels S-3 to S-5 (isotopic stage 5) and sapropels S-6 to S-8 (isotopic stage 7). Cyclicity is of the order of 20 000 yrs (precession cycles). (2) sapropels S-9, S-10, S-11 and S-12, each corresponding to a discrete isotopic stage. Cyclicity is of the order of 50 000 yrs (possibly axis inclination cycles).

5.9.   Irreversible effects of frequently repeated and long lasting anoxic episodes on benthic life

Stagnation was so severe and repetitive during the approximately 180 000 yrs encompassing stages 5 through 7 that sapropels actually represent 20–25% of the sediment coloumn. The duration of individual anoxic episodes for the most prominent sapropels, such as S-5 or S-6, is of the order of several thousand years. The effects of these repeated stagnations on benthic life were so destructive, that even now the population levels are much lower than they used to be in pre-stage 7 times, as resulting from studies in progress by E. Parisi (on KS 09) and A. Vismara Schilling (on Cobblestone cores from the western Mediterranean Ridge).

5.10.   Climatic significance of palynomorphs in sapropels

All the Late Quaternary sapropels from the eastern Mediterranean investigated by us yield amorphous organic matter of marine origin as a dominant constituent (in the range of 60–80%, see Table 4) with the exception of some resedimented levels of S-1 in Core Cobblestone 32. Organic matter contained in percentages <0.3% in normal, well-oxygenated sediments is prevailingly continental in origin (see highest and respectively lowest level analysed in Core Cobblestone 44).

Pollen grains recognized in sapropels are allochthonous to the marine realm, and are not responsible for the high organic carbon content. The very special geochemical environment of sapropels permits a good preservation of pollen grains and the study of the pollen record of the youngest sapropel permitted to document the transition from the late glacial—characterized by the steppe elements *Artemisia* and *Chenopodiaceae*—to the post-glacial, with temperate deciduous trees (Rossignol and Pastouret, 1971; Rossignol-Strick, 1975), a transition comparable to that recorded on land, in nearby Macedonia by Wijmstra (1969).

Pollen grains are basically airborne, but dispersal by turbidity currents was noticed in the Nile Cone province (Strick-Rossignol, 1973). Vegetational changes recorded in sapropel-bearing DSDP cores (Strick-Rossignol, 1973; Bertolani Marchetti and Accorsi, 1978) are climatically significant, but unfortunately they cannot be used to interpret the intriguing climatic record of the ice ages, because the latest Quaternary is never completely recorded in these cores.

Although detailed palynological studies were beyond the purposes of the present research, we point out a clear-cut discrimination between the isotopically and faunistically "warm" sapropel S-5, which is consistently devoid of bisaccate pollens of *Pinus*, and the "cold" sapropel S-6, where *Pinus* is a common record.

Palynology thus supports the other lines of evidence suggesting at least two possible mechanisms leading to stagnation, one of which is basically temperature-controlled, the other which is not.

6. Conclusion

Summarizing, Late Neogene Mediterranean sapropels seem to have originated in different ways. Most appear to be climatically modulated, but not all are temperature-dependent.

The highest percentage of organic carbon was recorded in Late Pliocene strata. Mean values resulting from several dozens of measurements on Late Pleistocene and Holocene sediments recovered during the Cobblestone Project (Cita *et al.*, 1982 and this paper) are 0.29% for normal sediments (marls and oozes) and 3.1% for sapropels.

Sapropels recording a definite cyclicity are found in isotopic stages 5 and 7 (cycle duration of the order of 20 000 years) and in stages 9 through 12 (cycle duration of the order of 60 000 years).

The older sapropel are poorly known; we believe that a great deal of work is still ahead of us before we fully understand how the eastern Mediterranean has become stagnant so many times in the last few million years of his history.

Acknowledgments

We thank the directors of AGIP Mineraria for permitting one of us (D.G.) to take part in this study. X-ray analysis presented in this paper were made at the Geochemical Laboratory of AGIP-SGEL. We gratefully acknowledge the helpful attitude of the director of the laboratory, Professor C. Morelli. We also thank the curator of the Deep Sea Drilling Project for granting our request of sapropel samples from Leg 42A. The Cobblestone Project was funded by CNR through grants 78.01878.66 and 79.02420.66; by NSF through grant OCE 77-20047. Discussions on Mediterranean stagnations with Bill Ryan, Floyd McCoy, Rob Thunell and Colette Vergnaud-Grazzini were of great help. Technical help was provided by R. Parisi, A. Malinverno and C. Broglia. We benefited from the critical reviews of A. Mörner, S. Schlanger and F. McCoy.

References

Anderson, R. V. 1933. The diatomaceous and fish-bearing Beida Stage of Algeria. *Journal of Geology* **4**, 673–698.

Arthur, M. A. 1979. Paleoceanographic events. Recognition, Resolution, and Reconsideration. *Review of Geophysics and Space Physics* **17** (7), 1474–1494.

Bennet, G. G. 1979. The Sedimentology, Diagenesis and paleo-oceanography of diatomites from the Miocene of Sicily. *Ph.d Thesis*, University of Durham.

Berger, W. H. 1976. Biogenous deep sea sediments: production, preservation and interpretation. *Chemical Oceanography* **5**, 265–388.

Bertolani Marchetti, D. & Accorsi, C. 1978. Palynological studies on samples from DSDP Leg 42A. *Initial Reports of the Deep Sea Drilling Project* **42A**, 789–804.

Bizon, G., Muller, C. & Vergnaud-Grazzini, C. 1979. Paleoenvironmental conditions during the deposition of diatomaceous sediments in Morocco and Cyprus. *Annales Géologyques des Pays Helleniques.* Hors série I, 113–128.

Bowen, D. G. 1978. *Quaternary Geology. A stratigraphic Framework for Multidisciplinary Work.* Cambridge: Pergamon Press, 1–221.

Blechschmidt, G., Cita, M. B., Mazzei, R. & Salvatorini, G. 1982. Stratigraphy of the western Mediterranean Ridge and southern Calabrian Ridge, Ionian Basin, eastern Mediterranean (U.S.A./Italy Cobblestone Project). *Marine Micropaleontology* 7, 101–134.

Casati, P., Bertozzi, P., Cita, M. B. Longinelli, A. & Damiani, E. 1978. Stratigraphy and paleoenvironment of the Messinian "Colombacci" Formation in the Periadriatic Trough. A Pilot Study. *Memorie della Società Geologica Italiana* **16**, 173–196.

Chamley, H., Dunoyer de Segonzac, G. & Mélières, F. 1978. Clay minerals in Messinian sediments of the Mediterranean area. *Initial Reports of the Deep Sea Drilling Project* **42A**, 389–395.

Ciaranfi, N. & Cita, M. B. 1973. Paleontological evidence of changes in the Pliocene climates. *Initial Reports of the Deep Sea Drilling Project* 13 (2), 1387–1399.

Cita, M. B. 1975. Planktonic foraminiferal zonation of the Mediterranean Pliocene deep-sea record: a revision. *Rivista Italiana di Paleontologia e Stratigrafia* 81 (4), 689–698.

Cita, M. B. 1979. Lacustrine and hypersaline deposits in the desiccated Mediterranean and their bearing on paleoenvironment and paleo-ecology. *Maurice Ewing Series* 3, American Geophysical Union, 402–419.

Cita, M. B., Broglia, C., Malinverno, A., Spezzibottiani, G. & Tomadin, L. 1982. Pelagic sedimentation in the southern Calabrian Ridge and western Mediterranean Ridge, eastern Mediterranean. *Marine Micropaleontology* 7, 135–162.

Cita, M. B., Chierici, M. A., Ciampo, G., Moncharmont Zei, M., d'Onofrio, S., Ryan, W. B. F. & Scorziello, R. 1973. The Quaternary Record in the Tyrrhenian and Ionian Basins of the Mediterranean. *Initial Reports of the Deep Sea Drilling Project* 13 (2), 1263–1339.

Cita, M. B. & Podenzani, M. 1980. Destructive Effects of Oxygen Starvation and Ash Falls on Benthic Life: a Pilot Study. *Quaternary Research* 13, 230–241.

Cita, M. B. & Ryan, W. B. F. 1973. The Pliocene record in deep-sea Mediterranean sediments. Time-scale and general synthesis. *Initial Reports of the Deep Sea Drilling Project* 13 (2), 1405–1416.

Cita, M. B. & Ryan, W. B. F. 1979. Late Neogene Environmental Evolution. *Initial Reports of the Deep Sea Drilling Project* **47A**, 447–460.

Cita, M. B., Vergnaud-Grazzini, C., Robert, C., Chamley, H., Ciaranfi, N. & d'Onorfio, S. 1977. Paleoclimatic record of a long deep-sea core from the eastern Mediterranean. *Quaternary Research* 8, 205–235.

Cita, M. B., Wright, R. A., Ryan, W. B. F. & Longinelli, A. 1978. Messinian Paleoenvironments. *Initial Reports of the Deep Sea Drilling Project* **42A**, 1003–1036.

Colalongo, M. L., Cremnini, G., Farabegoli, E., Sartori, R., Tampieri, R. & Tomadin, L. 1978. Paleoenvironmental study of the "Colombacci" Formation in Romagna (Italy): the Cella section. *Memorie della Società Geologica Italiana* **16**, 197–216.

Deroo, G., Herbin, J. P. & Rouchache, J. 1978. Organic geochemistry of some Neogene cores from Sites 374, 375, 377 and 378, Leg 42A, Eastern Mediterranean Sea. *Initial Reports of the Deep Sea Drilling Project* **42A**, 465–472.

d'Onofrio, S. & Iaccarino, S. 1978. Benthonic foraminifera from Italian Tortonian–Messinian sections. *International Meeting on the Geodynamic Effects of the Messinian Salinity Crisis in the Mediterranean,* Messinian Seminar No. 4 Abstracts.

Emiliani, C. 1955. Pleistocene temperature variations in the Mediterranean. *Quaternaria* 2, 87–98.

Erickson, A. J. & von Herzen, R. P. 1978. Down-hole temperature measurements, Deep Sea Drilling Project, Leg 42A. *Initial Reports of the Deep Sea Drilling Project* **42A**, 857–872.

Flint, R. F. 1957. *Glacial and Pleistocene Geology.* New York: Wiley, pp. 1–553.

Gersonde, R. 1978. Diatoms Paleoecology in the Mediterranean Messinian. *International Meeting on the Geodynamic and Biodynamic Effects of the Messinian Salinity Crisis in the Mediterranean.* Messinian Seminar, No. 4, Abstracts.

Gersonde, R. 1980. Palaeoekologische uns biostratigraphische Auswertung von Diatomeenassoziationen aus dem Messinium des Caltanissetta-Beckens (Sizilien) und einiger vergleichs-Profile in SO Spanien, NW Algerien und auf Kreta. *Dissertation,* University of Kiel, pp. 1–393.

Grosswald, M. G. 1980. Late Weichselian Ice Sheet of Northern Eurasia. *Quaternary Research* 13, 1–32.

Hajos, M. 1973. The Mediterranean Diatoms. *Initial Reports of the Deep Sea Drilling Project* 13 (2), 944–970.

Hieke, W., Sigl, W. & Fabricius, F. 1973. Morphological and structural aspects of the Mediterranean Ridge SW of Peloponnesus (Ionian Sea). *Bulletin of the Geological Society of Greece* 10, 109–126.

Hsü, K. J., Cita, M. B. & Ryan, W. B. F. 1973. The origin of the Mediterranean Evaporite. *Initial Reports of the Deep Sea Drilling Project* 13 (2), 1203–1232.

Hsü, K. J., Montadert, L. *et al.* 1978. *Initial Reports of the Deep Sea Drilling Project* **42A**, 1–1249.

Hsü, K. J., Montadert, L., Bernoulli, D., Cita, M. B., Erickson, A. J., Garrison, R. E., Kidd, R. B., Melières, F., Müller, C. & Wright, R. A. 1978. History of the Mediterranean Salinity Crisis. *Initial Reports of the Deep Sea Drilling Project* **42A**, 1053–1078.

Keigwin, L. D. & Thunell, R. C. 1979. Middle Pliocene climatic change in the western Mediterranean from faunal and Oxygen isotopic trends. *Nature* **282**, 294–296.

Keller, J., Ryan, W. B. F., Ninkovich, D. & Altherr, R. 1978. Explosive volcanic activity in the Mediterranean over the past 200,000 years as recorded in deep-sea sediments. *Geological Society of America Bulletin* **89**, 591–604.

Kidd, R. B., Cita, M. B. & Ryan, W. B. F. 1978. The stratigraphy of eastern Mediterranean sapropel sequences as recovered by DSDP Leg 42A and their paleoenvironmental significance. *Initial Reports of the Deep Sea Drilling Project* **42A**, 421–443.

Luz, B. & Bernstein, M. 1977. Planktonic foraminifera and quantitative paleoclimatology of the eastern Mediterranean. *Marine Micropaleontology* **1**, 307–323.

Maldonado, A. & Stanley, D. J. 1976. The Nile Cone: Submarine fan development by cyclic sedimentation. *Marine Geology* **20**, 27–40.

Martina, E., Casati, P., Cita, M. B., Gersonde, R., d'Onofrio, S. & Bossio, A. 1979. Notes on the Messinian Stratigraphy of the Crotone Basin, Calabria (Italy). *Annales Géologiques des Pays Helléniques*. Hors Série II, 755–765.

McCoy, F. W. 1974. Late Quaternary Sedimentation in the Eastern Mediterranean Sea. *Ph.D. Thesis*, Harvard University, 132pp.

McKenzie, J. A., Jenkyns, H. C. & Bennet, G. G. 1979. Stable isotopic study of the cyclic diatomite-claystones from the Tripoli Formation, Sicily: a prelude to the Messinian salinity crisis. *Paleogeography, Paleoclimatology, Paleoecology* **29** (1/2), 125–142.

Meulenkamp, J. E., Driever, B. W. M., Jonkers, H. A., Spaak, P., Zachariasse, W. J. & van der Zwaan, G. J. 1979. Late Miocene–Pliocene climatic fluctuations and marine "cyclic" sedimentation patterns. *Annales Géologiques des Pays Helléniques*. Hors Série II, 831–842.

Milliman, J. D. & Müller, J. 1973. Precipitation and lithification of magnesian calcite in the deep-sea sediments of the Eastern Mediterranean Sea. *Sedimentology* **20**, 29–45.

Mörner, N.-A. 1980. Ocean/land misfits and the 115,000 BP events. *International Geological Congress Abstracts II*, Paris.

Nesteroff, W. D. 1973. Petrography and mineralogy of sapropels. *Initial Reports of the Deep Sea Drilling Project* **13**, 713–720.

Olausson, E. 1961. Studies of deep-sea cores. *Reports of the Swedish Deep Sea Expedition 1947–48.* **8**(4), 323–438.

Olausson, E. 1965. Evidence of climatic changes in North Atlantic deep-sea cores with remarks on isotopic paleotemperature analysis. *Progress in Oceanography* **3**, 221–252.

Olson, E. A. & Broecker, W. S. 1959. Lamont natural radiocarbon measurements. VII. *American Journal of Sciences*, Radiocarbon Supplement **3**, 141.

Olsson, I. 1959. Uppsala natural radiocarbon measurements. I. *American Journal of Sciences*, Radiocarbon Supplement **1**, 87–102.

Parisi, E. & Cita, M. B. (in press). Late Quaternary paleoceanographic changes recorded by deep sea benthos in the western Mediterranean Ridge. *Geografia Fisica a Dinamica Quaternaria.*

Parker, F. L. 1958. Eastern Mediterranean Foraminifera. *Reports of the Swedish Deep Sea Expedition 1947–48* **8** (4), 217–283.

Ross, D., Neprochov, Y. P. *et al.* 1978. *Initial Reports of the Deep Sea Drilling Project* **42B**, 1–1244.

Rossignol, M. & Pastouret, L. 1971. Analyse pollinique de niveaux sapropélitiques post-glaciaires dans une carotte en Méditerranée orientale. *Révue de Paléobotanique et Palynologie* **11**, 227–238.

Rossignol-Strick, M. 1975. Palynologie des sapropels Méditerranéens du Villafranchien à l'Holocène. *Coll. Intern. CNRS* n. 129 *Méthodes quantitatifs et variations climatiques au cours du Pleistocène*, 93–102.

Rouchy, J.-M. 1980. La génèse des évaporites messiniennes de Méditerranée: un bilan. *Bulletin Centre Recherche Exploration—Production Elf-Aquitaine* **4** (1), 511–545.

Ruddiman, W. F. 1971. Pleistocene sedimentation in the equatorial Atlantic: Stratigraphy and Climatology. *Geological Society of American Bulletin* **82**, 283–302.

Ryan, W. B. F. 1969. The floor of the Mediterranean sea. *Ph. D. Thesis*, Columbia University, New York.

Ryan, W. B. F. 1972. Stratigraphy of late Quaternary sediments in the eastern Mediterranean. *The Mediterranean Sea* (Ed. D. J. Stanley), pp. 149–170.

Ryan, W. B. F. & Cita, M. B. 1977. Ignorance concerning episodes of ocean-wide stagnation. *Marine Geology* **23**, 197–215.

Ryan, W. B. F., Hsü, K. J. *et al.* 1973. *Initial Reports of the Deep Sea Drilling Project* **13**, 1–1453.

Schrader, H.-J. & Gersonde, R. 1978. The Late Messinian Mediterranean brackish to freshwater environment: diatom flora evidence. *Initial Reports of the Deep Sea Drilling Project* **42A**, 761–776.

Schrader, H. -J. & Matherne, A. 1981. Sapropel formation in the eastern Mediterranean Sea: evidence from preserved opal assemblages. *Micropaleontology* **27** (2), 191–203.

Shackleton, N. J. & Cita, M .B. 1979. Oxygen and carbon isotope stratigraphy of benthic foraminifers at Site 397: detailed history of climatic change during the Late Neogene. *Initial Reports of the Deep Sea Drilling Project* **47A**, 433–446.

Sigl, M., Chamley, H., Fabricius, F., Giround d'Argoud, G. & Müller, J. 1978. Sedimentology and environmental conditions of sapropels. *Initial Reports of the Deep Sea Drilling Project* **42A**, 445–465.

Sigl, W. & Müller, J. 1975. Identification and correlation of stagnation layers in cores from the eastern Mediterranean Sea. *Rapports Commission Internationale Mer Méditerranéenn* **23**, 277–279.

Stanley, D. J. & Blanpied, C. 1980. Late Quaternary water exchange between the eastern Mediterranean and the Black Sea. *Nature* **285**, 537–541.

Stanley, D. J. & Maldonado, A. 1977. Nile Cone: Late Quaternary stratigraphy and sediment dispersal. *Nature* **266**, 129–135.

Strick-Rossignol, M. 1973. Pollen analysis of some sapropel layers from the deep-sea floor of the eastern Mediterranean. *Initial Reports of the Deep Sea Drilling Project* **13** (2), 631–643.

Tchernia, P. & Lacombe, H. 1972. Caractères hydrologiques et circulation des eaux en Méditerranée. *The Mediterranean Sea* (Ed. D. J. Stanley), pp. 25–36.

Thunell, R. C. 1978. Distribution of Recent planktonic foraminifera in surface sediments of the Mediterranean Sea. *Marine Micropaleontology* **2**, 371–388.

Thunell, R. C. 1979 *a*. Climatic evolution of the Mediterranean Sea during the last 5.0 million years. *Sedimentary Geology* **23**, 67–79.

Thunell, R. C. 1979 *b*. Eastern Mediterranean Sea during the last Glacial Maximum: an 18,000 years BP reconstruction. *Quaternary Research* **11**, 353–372.

Thunell, R. C. & Lohmann, G. P. 1979. Planktonic foraminiferal fauna associated with eastern Mediterranean Quaternary stagnations. *Nature* **281**, 211–213.

Thunell, R. C., Williams, D. F. & Kennett, J. P. 1977. Late Quaternary paleoclimatology, stratigraphy and sapropel history in eastern Mediterranean deep-sea sediments. *Marine Micropaleontology* **2**, 371–388.

van Couvering, J., Berggren, W. A., Drake, R. E., Aguirre, E. & Curtis, G. H. 1976. The terminal Miocene event. *Marine Micropaleontology* **1**, 263–286.

van Straaten, L. M. J. W. 1972. Holocene Stage and Oxygen Depletion in Deep Waters of the Adriatic Sea. *The Mediterranean Sea* (Ed. D. J. Stanley), pp. 631–644.

Velitchko, A. A. 1977. Maximum glaciation in eastern Europe. The age problem. *IGCP Project 24: Quaternary Glaciations in the Northern Hemisphere*, Report No. 4.

Vergnaud-Grazzini, C. 1973. Etude écologique et isotopique de Foraminifères actuels et fossiles de Méditerranée. *Thèse*, Université de Paris.

Vergnaud-Grazzini, C., Cita, M. B. & Ryan, W. B. F. 1977. Stable isotopic fractionation, climate change and episodic stagnation in the eastern Mediterranean during the late Quaternary. *Marine Micropaleontology* **2**, 353–370.

Vismara Schilling, A., Stradner, H., Cita, M. B. & Gaetani, M. 1978. Stratigraphic investigations on the late Neogene of Corfou (Greece) with special reference to the Miocene/Pliocene boundary and to its geodynamic significance. *Memorie della Società Geologica Italiana* **16**, 279–318.

Wijmstra, T. A. 1969. Palynology of the first 30 meters of a 120 m deep section in northern Greece. *Acta Botanica Neerlandica* **18** (4), 511– .

Williams, D. F. & Thunell, R. C. 1979. Faunal and oxygen isotopic evidence of surface water salinity changes during sapropel formation in the eastern Mediterranean. *Sedimentary Geology* **23**, 81–93.

Williams, D. F., Thunell, R. C. & Kennett, J. P. 1978. Periodic fresh water flooding and stagnation of the eastern Mediterranean Sea during the Late Quaternary. *Science* **201**, 252–254.

10. Carbon-13 Cycle in Lake Greifen: A Model for Restricted Ocean Basins

J. A. McKenzie

Geology Institute, Swiss Federal Institute of Technology

Lake Greifen, a small lake in north-eastern Switzerland, has become progressively eutrophic during the last 100 years, as a consequence of higher rates of organic productivity facilitated by an increased nutrient supply. The oxidation of the excess organic matter causes a severe depletion of oxygen in the bottom waters producing anoxic conditions in midsummer. The historical process of eutrophication is traced using changes in the carbon-13 content of the lacustrine chalks, which are mostly composed of calcite precipitated in the surface waters. At first, increased productivity resulted in a carbon-13 depletion in the chalks because the excess organic matter, which is depleted in carbon-13, was oxidized and recycled in the lake. With further, more extreme productivity increase and subsequent anoxicity in the bottom waters, the excess organic matter was stored in the sediments, and the carbon-13 content of the chalk showed an enrichment. This stable isotope study is presented as a model to explain the carbon-13 enrichment in pelagic carbonates during the Middle and Early Cretaceous.

1. Introduction

Major carbon-13 increases in the isotopic composition of Middle and Lower Cretaceous pelagic carbonates have been tentatively correlated with the widespread deposition of black shales (Weissert *et al.*, 1979 and Scholle & Arthur, 1980). Changes in circulation patterns and/or increased productivity in surface waters are two mechanisms which have been proposed to explain this isotopic increase. The storage of significant quantities of carbon-13-depleted organic matter in anoxic sediments could also increase the carbon-13 content of the oceanic bicarbonate. The relative importance of these mechanisms remains controversial. In this paper, an isotopic study of carbonate sediments and dissolved bicarbonate in an eutrophic lake is presented as an actualistic model for carbon-13 distribution in an ocean basin during times of increased productivity and widespread anoxia.

Lakes are natural "beakers" in which geochemical processes can be effectively studied. In general, they are much quicker to respond to environmental pressures than ocean basins, and, because of the smaller size of the reservoirs, the geochemical signals of such perturbations is amplified. The carbonate sediments and waters of Lake Greifen, a highly productive, eutrophic lake in north-eastern Switzerland (Figure 1), provide an opportunity to study changes in carbon-isotope abundances resulting from changes in productivity. During the last 100 years, agricultural and industrial activity around the lake augmented the normal nutrient influx with excess phosphates and nitrates, which facilitated an increase in organic productivity. Yearly, in midsummer, the oxygen content of the bottom waters is severely depleted by the oxidation of the excess organic material. This eutrophication process is reflected in the lacustrine sediments which have progressively changed from totally oxygenated to anoxic in the last 100 years.

The lacustrine sediments are composed predominately of calcite which was

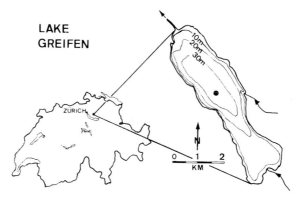

Figure 1. Location map for Lake Greifen, Switzerland. The coring and water-sampling site are indicated
     by the dot marking the deepest point in the lake.

precipitated inorganically in the surface waters as a result of $HCO_3^- - CO_3^{2-}$
disequilibrium during periods of peak biological activity (Kelts & Hsü, 1978). The
lacustrine calcite incorporates dissolved bicarbonate from the surface waters. A
study of calcite and dissolved bicarbonate collected simultaneously from Lake
Greifen indicates that the calcite is precipitated in apparent carbon isotopic
equilibrium with the bicarbonate (McKenzie, unpublished data). Core samples
should, therefore, faithfully record any carbon isotopic change that occurred as a
result of the historical trend towards increased algal productivity paralleling the
progressive eutrophication of the lake. Furthermore, the annual winter overturn and
subsequent oxygenation of the deeper waters allow one to follow the evolution of the
carbon-13 content of the dissolved bicarbonate throughout the water column with
the summer's biological activity in the surface waters and subsequent return to
anoxic conditions in the deeper waters.

2. Description of lacustrine chalk in core

Short gravity cores were taken from the deepest point (32 m) near the center of the
lake (Figure 1). The least disturbed core (67 cm long, was chosen for this study. The
stratigraphy of the core is shown in Figure 2. The top 25 cm comprise varve couplets
which consist of white to pale yellow calcareous layers alternating with black,
gelatinous layers, which are rich in organic material and diatoms. The freshly-cut
surface of this interval is coal black reflecting the abundance of unstable iron sulfides
formed under reducing conditions. Bottom-water anoxia throughout most of the
year prevents bioturbation and preserves the varves. The organic-carbon content
ranges from 3 to 8% (Giger et al., 1980). The base of the varve section is transitional.
It grades down into a yellow–gray marl with progressively fewer black sulfide
streaks. Laminations are faint and commonly disturbed by burrowing. At 42 cm, the
sediment becomes a rather homogenous, bioturbated, compact yellow-gray marl.
This transitional interval from laminated to homogenous sediments (25–42 cm)
represents the progressive diminishment of eutrophic conditions in the lake. Below
42 cm to the base at 67 cm, the sediments appear to have been deposited in
completely oxygenated waters. At 47 cm, there is a 2 mm thick, reddish-brown
turbidite horizon produced by a nearshore slump.
     Sedimentation rates ranged from 3–8 mm yr^{-1} based on ^{210}Pb measurements and
the counting of annual varve couplets (Gäggeler et al., 1976; Emerson & Widmer,

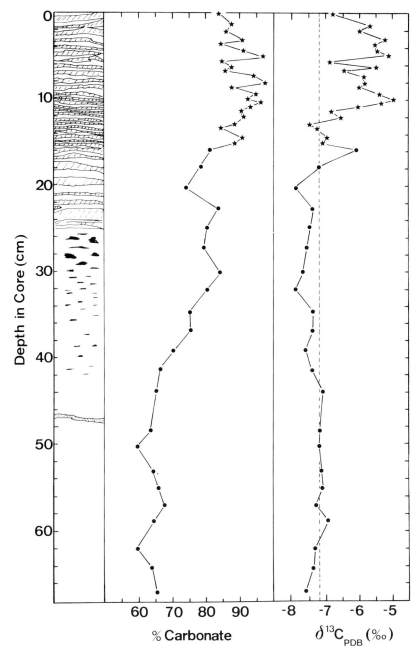

Figure 2. The percentage of carbonate and carbon-isotope stratigraphy of the Lake Greifen short core. The progressive eutrophication of the lake is depicted by the change from homogenous marls to varved sediments, white chalks alternating with black organic- and diatom-rich layers. With the increasing bottom water anoxia, the sediments are pigmented by black, unstable iron sulfides. Data for samples taken over 0.5 cm intervals are represented by dots, while those from individual chalk varves are represented by stars. The dashed line represents the average $\delta^{13}C$ value $(-7.1\ \%_{oo})$ for pre-eutrophic lacustrine chalk. See text for further discussion.

1978). The highest rates are recorded for the upper varved section and decrease gradually downward. Since the dry sediment mass accumulation rate of the varves remains relatively constant, the decrease in sedimentation rate is assumed to be an artifact of compaction (Emerson and Widmer, 1978). Below 25 cm, the sedimentation rate is estimated to have averaged 3 mm per yr. Using these estimated rates, the base of the varved section is placed at 50 years BP, the onset of the progressive eutrophication at approximately 100 years BP and the base of the core at approximately 190 years BP.

3. Experimental methods

3.1. Lacustrine calcite

The opened core halves were allowed to air dry completely to facilitate the separation of individual varves. The calcareous members of the first 25 couplets were separated and analysed. Below 15 cm, the core was sampled every 2 cm over a 0.5 cm interval. All samples were treated with a 6% sodium hypochloride solution to remove organic material. The relative weight percentage of calcite and dolomite was estimated from x-ray diffraction patterns and the total weight percentage of carbonate was determined using a combustion-titration apparatus. Stable isotope analyses of CO_2 gas released from the samples by the traditional phosphoric acid method (McCrea, 1950) were made with a triple collecting mass spectrometer, Micromass 903. The carbon isotope ratio of the lacustrine chalk is reported as the per mil (‰) deviation from the PDB isotopic standard. Appropriate correction factors have been included (Craig, 1957).

Clastic sources to the lacustrine sediment have a relatively constant ratio of dolomite and calcite, which I used to correct the isotope results for small amounts of detrital calcite in the chalk sediments. It is assumed that all dolomite is detrital. Therefore, after the dolomite–calcite ratio and the isotopic content of the fine-grained clastic input were determined from the present river influx and from the carbonate rocks in the catchment area of the lake, the percentage of dolomite was used to estimate the proportion of detrital calcite in each sample; thus, the carbon-isotope ratios are adjusted to a pure lacustrine chalk value, which is always more negative than the measured value. This correction filters any isotopic changes due to variations of detrital input that may have occurred as a result of man's activities during the past 200 years. The corrected carbon-13 values of the pure lacustrine chalk and the percentage of carbonate content are graphically shown in Figure 2.

3.2. Dissolved bicarbonate

Every four weeks at the same hour, water samples were collected using a Nansen bottle from a profile at the deepest part of Lake Greifen, as well as from the two major inflow rivers and the outflow river (Figure 1). The water was filtered through analytical-grade filter paper to remove particulate matter, and a concentrated NaOH solution was added to raise the pH to greater than 10. A saturated $BaCl_2$ solution is added to fix all of the dissolved bicarbonate as a $BaCO_3$ precipitate. Carbon-isotope ratios were measured on the CO_2 gas released from the $BaCO_3$ precipitate, which had been previously treated with a 6% sodium hypochloride solution. The plot of carbon-13 values of the dissolved bicarbonate from various depths for monthly profiles are shown in Figure 3.

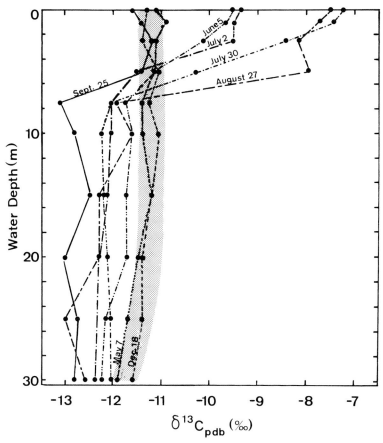

Figure 3. Monthly δ^{13}C profiles of the dissolved bicarbonate in Lake Greifen. The carbon-13 increase in
surface waters due to photosynthesis and the carbon-13 depletion in deep waters resulting from the
oxidation of sinking organic material are depicted. The shaded area represents the range of δ^{13}C
values found between December and May.

4. Results and discussion

A significant rise in the percentage of carbonate (Figure 2) from an average of 64% for
oxygenated sediments to up to 98% in individual varves is apparently directly related
to the eutrophication of Lake Greifen. The increased nutrient input resulting from
man's activities around the lake stimulated an increase in biological productivity.
The extraction of CO_2 during photosynthesis by phytoplankton displaces the
carbonate equilibria in the surface waters and results in the inorganic precipitation of
calcite (Kelts and Hsü, 1978). Therefore, in general, the greater the biological
activity, the more calcite will be precipitated. Carbon-isotope data from coexisting
calcite precipitate and dissolved bicarbonate indicate that the lacustrine chalk is
precipitated in isotopic equilibrium. Therefore, the changes in the δ^{13}C value of the
chalk can be directly related to changes in the δ^{13}C value of the dissolved bicarbonate,
which in turn is related to changes in the degree of biological activity. Since there is
no indication that circulation patterns in the lake have changed over the last 200
years, δ^{13}C fluctuations can be interpreted as changes in the magnitude of
productivity.

During photosynthesis, organisms preferentially concentrate the lighter isotopes (carbon-12) into organic matter leaving the bicarbonate reservoir enriched in the heavier isotope (carbon-13). This trend is clearly illustrated by the $\delta^{13}C$ values of the dissolved bicarbonate from the monthly profiles (Figure 3). From December to May, the $\delta^{13}C$ value remains relatively constant with depth in spite of a near-surface temperature rise of 15°C from February to May. With the onset of the summer growth period, biological activity enriches the surface waters in carbon-13 by approximately 4‰. For comparison, the $\delta^{13}C$ value of the bicarbonate entering the lake during the same period changes less than 1‰, while the $\delta^{13}C$ value of the bicarbonate in the outflowing water shows exactly the same carbon-isotopic pattern as recorded throughout the year in the near-surface waters. The actual carbon-13 enrichment in the surface waters should be greater than the apparent enrichment due to the continued inflow of waters with $\delta^{13}C$ values of -10.5‰ and -12‰ (average values for the two major inflow rivers). Biweekly, sediment-trap collections of sediment settling out of the water column show that the maximum amount is sedimented during the summer months, the time of highest productivity (Weber, 1981). Predictably, the sediment-trap chalks from the summer months are enriched in carbon-13 corresponding to the simultaneous carbon-13 enrichment in the dissolved bicarbonate (McKenzie, unpublished data).

Below 5 m water depth, the decomposition of the sinking organic material releases CO_2 which decreases the $\delta^{13}C$ value of the dissolved bicarbonate and depletes the oxygen content of the underlying waters. The extreme eutrophic conditions in Lake Greifen result in an anaerobic hypolimion during most of the year (Pleisch et al., 1972, 1975). The annual winter overturn of the lake returns oxygen to the bottom waters and once again equalizes the $\delta^{13}C$ value of the dissolved bicarbonate from top to bottom. The yearly replenishment of oxygen in the bottom waters is not sufficient to affect the anoxic conditions in the sediments, which remain undisturbed by bioturbation.

Although more complex, the $\delta^{13}C$ trends in the marine environment are analogous to those measured for the Lake Greifen. Numerous profiles of the oxygen concentration and $\delta^{13}C$ ratios from the Atlantic and Pacific Oceans document the parallel consumption of oxygen in the intermediate waters by the decomposition of carbon-13-depleted organic matter with decreases in the $\delta^{13}C$ ratio of the dissolved bicarbonate (Deuser and Hunt, 1969; Duplessy, 1972; Kroopnick, 1974). In the surface waters, the preferential incorporation of isotopically-light carbon into organic material during photosynthesis, as well as atmospheric exchange and circulation patterns, determines the $\delta^{13}C$ value of the dissolved CO_2 (Kroopnick et al., 1977).

This isotopic study of the chalk in the Lake Greifen core establishes the effect of increased productivity on the $\delta^{13}C$ value of the precipitated calcite. The average $\delta^{13}C$ value for the lacustrine chalk in the homogenous, oxygenated sediments between 42 and 67 cm is -7.1 (±0.2) ‰ and is taken as the normal value for the pre-eutrophication calcite (Figure 2). Between 20 and 42 cm, the $\delta^{13}C$ value of the calcite shifts slightly in a negative direction with an average value of -7.5 (±0.2) ‰. The lacustrine chalk from this interval corresponds to the time of progressive eutrophication of the lake ending in laminated sedimentation. The slight decrease in the carbon-13 content of the chalk can be adequately explained by the addition of carbon-13-depleted CO_2 from the decomposition of the excess organic material resulting from the augmented productivity. The carbon-13-depleted CO_2 released in the deeper waters and eventually refluxed upward into the near-surface waters is a major source of carbonate for calcite precipitation (Kelts & Hsü, 1978). During the phase of progressive increase in productivity, the bottom waters of the lake continued to be capable of oxidizing the greater part of the additional organic matter, and the organic CO_2 was recycled into the lake's chemical system. Hence, the $\delta^{13}C$ value of the lacustrine chalks decreases.

Within the last approximately 40 years (0–20 cm), a critical boundary was crossed. The lake has become extremely eutrophic, producing anoxic bottom water conditions during the summer months; since 1941, the oxygen concentration drops in summer to less than 1 mg l^{-1} in waters below 10 m (Thomas, 1955; Pleisch et al., 1972, 1975). The oxygen content of the waters is incapable of handling increased input of organic material. Each summer the oxygen is quickly exhausted and organic material settles to the bottom, where the carbon-13-depleted carbon is preserved in the reduced state. In contrast to earlier times when organic CO_2 was refluxed, the continued removal and storage of the reduced carbon enriches the surface water bicarbonate in carbon-13. An increase in the $\delta^{13}C$ value of the lacustrine chalks during the eutrophic period documents this enrichment (Figure 2). Although the measurement of individual varves provides a high resolution of the yearly isotopic changes, the overall trend shows that the varved chalk (0–20 cm) has greater $\delta^{13}C$ values than the homogenous chalk from the oxygenated zone (42–67 cm). Basically, there is a positive correlation between the yearly fluctuations in the percentage of carbonate of the varves and the $\delta^{13}C$ value of the chalk; the samples with the highest percentage of carbonate tend to be the most enriched in carbon-13. This observation is in accordance with the postulate that at times of greatest productivity more calcite is precipitated and this calcite is more enriched in carbon-13 due to the preferential removal of carbon-12 during photosynthesis.

5.   Lake model for carbon-13 cycle

The progressive eutrophication of Lake Greifen can serve as a model for carbon-13 fractionation within an ocean basin during periods of changing fertility patterns. The model, schematically depicted in Figure 4, implies that increases in biological productivity in conjunction with the active return of intermediate waters to the surface should result in a slight carbon-13 depletion in the dissolved CO_2 of the upper waters. The additional organic material is not preserved but is oxidized and refluxed back into the surface waters, where it can be utilized by carbonate-secreting planktonic organisms. Hence, the $\delta^{13}C$ value of their tests will be more depleted in carbon-13 than prior to the time of increased fertility. This case models an area of oceanic upwelling.

A major, more extensive, productivity increase in a stagnant or sluggish ocean will have the opposite effect on the $\delta^{13}C$ value of the pelagic carbonates. Because of restricted circulation, oxygen replenishment occurs at rates slower than oxygen consumption leading to anoxic bottom waters. Surplus organic material reaches the ocean floor and is preserved in the anoxic sediments. As in the eutrophic Lake Greifen, the constant removal and storage of the carbon-13 depleted matter from the surface waters leaves the dissolved CO_2 enriched in carbon-13. As a consequence, tests of carbonate-secreting planktonic organisms will be enriched in carbon-13. Oxygenated and anoxic pelagic carbonate sediments throughout the ocean basin should show a positive carbon-13 shift during these times of higher productivity. This case models a restricted ocean basin.

Although the limnology of Lake Greifen is not equivalent to the oceanography of ocean basins, important considerations evolve from this simple model for carbon-13 distribution. The lake model analogy suggests that carbon-13 changes in the ancient marine record could be dependent upon changes in oceanic productivity. In the marine environment, the effects of changes in circulation and fertility patterns upon the carbon-13 cycle are not readily separated. The two mechanisms are definitely

INCREASED PRODUCTIVITY

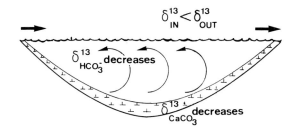

~with CO₂ reflux

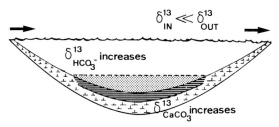

~with stagnation

Figure 4. The lake model for carbon-13 cycle in an ocean basin during times of increased productivity. In a basin where the additional organic material is oxidized and refluxed into the surface waters, the $\delta^{13}C$ value of the carbonate sediments decreases. In a basin with sluggish circulation, the bottom-water oxygen is consumed and the additional organic material is preserved in the sediments. An increase in the $\delta^{13}C$ value of the carbonate sediments reflects this removal of carbon-13 depleted material from the system.

interrelated. The uniqueness of the lake model developed from the Lake Greifen study is that the effect of productivity upon the carbon-13 cycle in the lake could be isolated and studied. A similar example does exist in the ancient marine record. At the Cretaceous–Tertiary boundary, there was a geologically rapid change in oceanic productivity due to the massive extinction of planktonic fauna and flora. As summarized in Scholle and Arthur (1980), the $\delta^{13}C$ ratio of pelagic carbonates from numerous Cretaceous–Tertiary boundary sections abruptly decreases and subsequently increases. In a recent study (Hsü et al., 1982), a detailed stable isotopic analysis of an extended Cretaceous–Tertiary boundary section from the South Atlantic, DSDP Site 524, demonstrated that the carbon-13 depletion in carbonate sediments following the "Cretaceous Terminal Event" was not instantaneous but was progressive and occurred over a period of 30 000 to 40 000 years. After reaching a minimum value, the $\delta^{13}C$ ratio returned to a pre-Tertiary value in about 300 000 to 400 000 years after the boundary event. Hsü et al. (1982) interpret the carbon-13 depletion in the sediment as a consequence of the massive decrease in oceanic photosynthesis resulting in an increase in the carbon-12 content of the dissolved CO_2. The turnabout in the $\delta^{13}C$ profile at 40 000 years after the boundary event reflects the progressive return to more normal oceanic productivity.

Numerous positive carbon-13 shifts have been recorded in Lower and Middle Cretaceous carbonates and have been correlated with the occurrence of widespread anoxic sediments and the burial of organic matter (Scholle & Arthur, 1980). Although the oceanographic factors are extremely complex, it is interesting to speculate the influence of increased productivity in causing these positive carbon-13 shifts and the applicability of the lake model. What is the relationship between anoxic sediments and productivity? Is it simply decreased circulation which produces the anoxic conditions necessary for the preservation of organic matter or is it a concurrent increased productivity which promotes the occurrence of anoxic bottom waters? The lake model interpretation would suggest that the positive carbon-13 shifts reflect increased productivity which, in turn, helps to facilitate the anoxia.

A detailed study of the relationship between shale deposition in a restricted ocean basin and positive carbon-13 shifts was presented by Weissert et al. (1979) and Weissert & McKenzie (1981). They studied the cyclic pelagic sediments of the southern Alpine Maiolica Formation, which were deposited in the Lower Creta-ceous (Barremian) Lombardian Basin, a part of the western Tethys Ocean. In these sediments, the $\delta^{13}C$ value of the calcium carbonate fraction of black shales was essentially more positive than those for the surrounding white nannofossil limes-tones. Cyclic, carbon-13 increases were also found in the nannofossil limestones of the adjacent Trento Plateau, where black shales of the same time period are missing. They interpreted the cyclic carbon-13 increases as a result of changes in circulation due to climatic changes; a thermohaline circulation system comparable with the pattern in the Mediterranean Sea was replaced by a stratified ocean during more humid periods of the Barremian. Changes in productivity were discounted as a mechanism for the carbon-13 increases, but the data can be further interpreted in light of the lake model, as shown below.

In the case of Lake Greifen, increased nutrient input resulted in increased production and anoxic bottom waters. During more humid periods in Barremian times, increased runoff into the western Tethys Ocean could have provided the increased nutrient influx required to augment productivity. With the changeover from thermohaline circulation to basinal stagnation, productivity increased causing an increase in the carbon-13 ratio of dissolved bicarbonate plus enhancing anoxic conditions in the bottom waters. Decreased productivity should then occur during the reverse transformation from stagnant conditions to thermohaline circulation. The increase or decrease of carbon-13 content of the surface waters occurred throughout the basin as isotopically recorded in both basin and plateau sequences.

Figure 5 is a sketch of a typical Maiolica transition from a finely laminated black shale through laminated to bioturbated gray limestone to a homogeneous white limestone. This transition shows a progressive $\delta^{13}C$ decrease from $+2.7$ to $+1.3\%_0$.

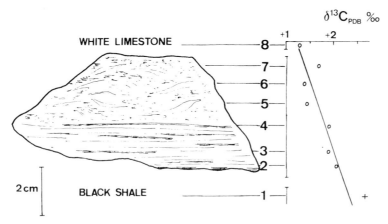

Figure 5. Sketch of a typical transition from black shale through gray laminated to bioturbated limestone to white homogenous limestone. Points 1 to 8 represent $\delta^{13}C$ values for a series of samples taken from the Suello profile of the southern Alpine Maiolica Formation.

The linear decrease in the $\delta^{13}C$ value from black shale to white limestone could be explained as a result of decreasing productivity across this interval. Perhaps, the linear carbon-isotopic changes reflect the progressive influence of productivity changes on the carbon-isotopic composition of the surface waters. Superimposed upon these productivity alterations are the factors resulting from changes in circulation patterns. In the Maiolica Formation, the black shale–white limestone end-members represent the result of major circulation changes, i.e. from an anoxic to an oxygenated ocean basin. The carbon-isotopic transition between these two end-members represents the effect of changing climatic conditions on productivity.

One consequence of a stagnant ocean basin is the insufficient replenishment of the nutrient content of the surface waters because of the diminished upwelling of underlying nutrient-rich waters. If productivity is to remain at an increased level during times of black shale formation, a renewed supply of nutrients is essential. The continued depletion of surface-water nutrients will result in reduced productivity. The analogy to the eutrophic lake model could provide an answer to this apparent dilemma of productivity versus nutrient supply. In Lake Greifen, the annual, winter overturn brings nutrients to the surface waters. The oxygen content of the bottom waters is also renewed but quickly consumed by the decomposition of sinking organic matter photosynthesized during the annual diatom and algal blooms in the overlying waters. The return to anoxia is rapid enough to prevent the repopulation by bottom dwellers, and the sediments maintain their undisturbed, laminated character. The frequency and extent of overturn in an ocean basin is different than in a lake, but the effect could be very similar. The overturn of stagnant waters brings nutrient-rich waters to the surface and blooms occur. Renewed productivity accelerates the return to bottom water anoxia with a rapidity sufficient to prevent the overturn from being recorded in the sedimentary record. This productivity also reinforces the positive shift in the carbon-13 value of the pelagic carbonates.

Although the lake model may not be universally applicable to major ocean basins, it can be utilized to evaluate the effect of productivity changes upon the carbon-13 cycle. The model demonstrates that productivity increases or decreases are sufficient to alter the $\delta^{13}C$ ratio of the dissolved CO_2 in surface waters. In the case of restricted, marginal ocean basins, as the southern Alpine Lombardian Basin or even the Middle Cretaceous Atlantic system, changes in circulation patterns complicate a strict productivity interpretation for the carbon-13 shifts, but, nevertheless, the major effect of productivity cannot be ignored.

Acknowledgments

I gratefully acknowledge K. Hsü, K. Kelts and H. Weissert for their stimulating discussions and M. Arthur, W. Giger and H. Jenkyns for their thoughtful reviews of the manuscript. I give special thanks to G. Eberli, J. Pika, H. P. Weber and J. Weiss for their invaluable assistance with the water sampling, coring and data collecting. This work was supported by the Swiss National Science Foundation (Limnogeology Project) and is Contribution No. 166 of the Laboratory for Experimental Geology, E.T.H., Zürich.

References

Craig, H. 1957. Isotopic standards of carbon and oxygen and correction factor for mass-spectrometer analysis of carbon dioxide. *Geochimica et Cosmochimica Acta* **12**, 133–149.
Deuser, W. G. & Hunt, J. M. 1969. Stable isotope ratios of dissolved inorganic carbon in the Atlantic. *Deep Sea Research* **16**, 221–225.

Duplessy, J.-C. 1972. La Geochimie des Isotopes Stables du Carbone dans la Mer. *Unpublished Ph.D. dissertation*, L'Universite de Paris VI, Paris.

Emerson, S. & Widmer, G. 1978. Early diagenesis in anaerobic lake sediments—II. Thermodynamic and kinetic factors controlling the formation of iron phosphate. *Geochimica et Cosmochimica Acta* **42**, 1307–1316.

Gäggeler, H., von Gunten, H. R. & Nyffeler, K. 1976. Determination of ^{210}Pb in lake sediments and in air samples by direct gamma-ray measurement. *Earth and Planetary Science Letters* **33**, 119–121.

Giger, W., Schaffner, C. & Wakeham, S. G. 1980. Aliphatic and olefinic hydrocarbons in recent sediments of Greifensee, Switzerland. *Geochimica et Cosmochimica Acta* **44**, 119–129.

Hsü, K. J., He, Q., McKenzie, J. A. *et al.*, 1982. Mass mortality and its environmental and evolutionary consequences. *Science* **216**, 249–256.

Kelts, K. & Hsü, K. 1978. Freshwater carbonate sedimentation. *Lakes: Chemistry, Geology, Physics*. (Ed. Lerman, A.) New York: Springer-Verlag, pp. 295–323.

Kroopnick, P. 1974. The dissolved O_2—CO_2—^{13}C system in the eastern equatorial Pacific. *Deep Sea Research* **21**, 211–227.

Kroopnick, P. M., Margolis, S. V. & Wong, C. S. 1977. δ^{13}C variations in marine carbonate sediments as indicators of the CO_2 balance between the atmosphere and oceans. *The Fate of Fossil Fuel CO_2 in the Oceans*. (Eds Andersen, N. R. & Malahoff, A.) New York: Plenum, pp. 295–321.

McCrea, J. M. 1950. The isotopic chemistry of carbonates and a paleotemperature scale. *Journal of Chemical Physics* **18**, 849–857.

Pleisch, P., Mattenberger, H. A. & Peyer, A. 1972, 1975. Dokumentation "Greifensee Wasser". Verband zum Schultze des Greifensee. Uster, Switzerland.

Scholle, P. A. & Arthur, M. A. 1980. Carbon isotope fluctuations in Cretaceous pelagic limestones: potential stratigraphic and petroleum exploration tool. *The American Association of Petroleum Geologists Bulletin* **64**, 67–87.

Thomas, E. A. 1955. Stoffhaushalt und Sedimentation in Oligotrophen Aegerisee und im eutrophen Pfäffiker- und Greifensee. *Memorie dell'Istituto Italiano di Idrobiologia* **8**, 357–465.

Weber, H. P. 1981. Sedimentologische und Geochemische Untersuchungen im Greifensee (Kanton Zürich/Schweiz). *Unpublished Ph.D. dissertation*, Swiss Federal Institute of Technology, Zürich.

Weissert, H., McKenzie, J. & Hochuli, P. 1979. Cyclic anoxic events in the Early Cretaceous Tethys Ocean. *Geology* **7**, 147–151.

Weissert, H. J. & McKenzie, J. A. 1981. Anoxic sediments in the early Cretaceous Tethys Ocean: an oceanographic explanation. *IAS 2nd European Meeting Bologna, Abstracts*, 205–206.

11. Progressive Ventilation of the Oceans— Potential for Return to Anoxic Conditions in the Post-Paleozoic

P. Wilde

Lawrence Berkeley Laboratory, University of California

W. B. N. Berry

Department of Paleontology, University of California

After the ventilation of the residual anoxic layer in the Late Paleozoic a return to ephemeral anoxic conditions in the ocean is suggested by anoxic sediments found at numerous Deep Sea Drilling Project sites in cores of Cretaceous age. Such conditions would be similar to that in the CO_2—H_2S zones described by Borchert (1965) for marine iron-ore formation. A preliminary physical oceanographic model is presented to explain the development of oxygen-depleted layers in mid-waters below the surface wind-mixed layer during non-glacial climates. The model shows the range of temperature, salinity and density values for hypothetical water masses for two climatically related oceanographic situations: Case A where bottom waters are formed at mid-latitudes at the surface salinity maxima, and Case B where bottom waters are produced at high latitudes but not by sea-ice formation as in the modern ocean. The hypothetical water masses are characterized by examples from the modern ocean and extrapolation to non-glacial times is made by eliminating water masses produced by, or influenced by, sea-ice formation in modern glacial times. The state of oxidation is determined by plotting the model water masses on an oxygen saturation diagram and comparing the relative oxygen capacity with modern conditions of zonal organic productivity. The model indicates for Case A (high latitude temperatures above 5°C) two oxygen-depleted layers in the equatorial regions: (1) from about 200 m to the depth of complete oxidation of surface material separated by an oxygenated zone; (2) a deep depleted zone along the base of the pycnocline at 2900 m. The deep depleted zone extends along the Case A pycnocline polarward toward the high latitude productivity maximum. For Case B with a pycnocline at about 1500 m the deep anoxic layer is not sustained. Considerations of density only suggest that neutral stratification and the potential for overturn is enhanced for climates transitional between Case A and Case B where the density contrast between major water masses formed at high latitudes and mid-latitudes is minimal or non-existent.

1. Introduction

Berry and Wilde (1978) noted that dark (commonly black), organic carbon-rich shales and mudstones are present in relatively greater amounts and are spread more widely geographically in the Lower Paleozoic stratigraphic record than they are in the rock record of younger geologic intervals. As many of the Lower Paleozoic black shales formed in continental slope and ocean basinal settings, Berry and Wilde (1978) suggested that the areally widespread aspect of these coeval Lower Paleozoic black shales might reflect the presence of a residual anoxic layer in the open oceans which was eventually ventilated by the end of the Paleozoic.

The model of progressive open ocean ventilation (Berry & Wilde, 1978), implies that the early earth's ocean was anoxic. As soon as oxygen was present in the atmosphere, after the development of photosynthesis (Cloud, 1976), oxygen would have been stirred into the surface waters of the oceans by wind-mixing. Wind-mixing doubtless resulted in a ventilated upper or mixed layer of the ocean. Ventilation of deep ocean waters, reasoning by analogy with modern processes, would have resulted primarily from formation of cold, oxygen-rich waters at high latitudes during periods of glaciation or at least of sea-ice formation there. Cold, oxygen-laden waters sank and spread towards the equator through the deep oceans. Sustained sinking of oxygen-rich waters would have oxygenated the deep oceans. Then, by advective upward mixing, the mid-waters, those between the mixed layer and the deep ocean waters, would have been ventilated progressively. Thus, a residual parcel of mid-ocean water that was anoxic would have been ventilated gradually from below. That ventilation process would have been countered to a degree by a loss of oxygen resulting from the decay and decomposition (oxidation) of organisms that lived in the surface and sank after death. As the residual parcel of anoxic mid-water was gradually ventilated, it could have become anoxic again, even after being ventilated (see Berry & Wilde, 1978, Figure 3). Intervals of decreased oxygen supply, either from reduced deep ocean circulation and advective mixing, or from increased oxygen "consumption" following from a high level of organic productivity at the surface and decay of those organisms, would have tended to make that parcel of mid-water with low oxygen content anoxic.

Whereas Berry and Wilde (1978) drew attention to the ventilating potential inherent in cold, polar waters that bear relatively large quantities of oxygen, a second ventilating mechanism may also be brought into consideration. Chamberlain (1906) and Arthur and Natland (1979) noted that warm but relatively saline waters are denser than most surface water and may sink to some density equilibrium depth. Warm but relatively salty water formed in the present-day Mediterranean and Sargasso Seas, where evaporation exceeds precipitation, results in denser, oxygenated waters that sink and mix with other waters in the deep oceans. During non-glacial geologic intervals warm, equable climates were widespread and oceanic circulation was sluggish. In the regions where evaporation exceeded precipitation, relatively dense but saline and oxygenated water could form to produce deep or even bottom water in the absence of cold, high latitude deep water. Areas where evaporation exceeds precipitation today are centered at approximately 30° North and South latitude, at the latitudes of major deserts on land.

Potentially, a certain degree of ocean ventilation could have resulted from salinity driven sinking of oxygenated water. Because warm water contains less oxygen than cold water, ventilation by salinity driven sinking would not have been as effective as that from cold, dense water. In the model presented herein, both ventilation mechanisms are considered and the consequences of mixing cold, dense and saline, dense waters are described.

2. The stratigraphic record of potential ocean basin and slope sediments

Lower Paleozoic rock suites that formed on probable continental slopes and in ocean basins in the residual anoxic layer commonly are black, organic-rich shales, mudstones, and limestones (Berry & Wilde, 1978). Such rock suites of Cambrian through Devonian age have been described in the following: Berry (1974), Berry & Boucot (1968, 1973), Cook & Taylor (1977), Cowie (1974), Dean (1980), Glaeser (1979), Heckel & Witzke (1979), Holland (1971), Krebs (1979), Leggett (1980), Martinsson (1974), Rushton (1974) and Stewart & Poole (1974).

Stratigraphic records of the Late Paleozoic and Mesozoic shelf and ocean basin sequences (Bernouilli & Jenkyns, 1974; Jenkyns, 1978, 1980; Ross, 1979; Stewart *et al.*, 1977) suggest that ocean ventilation had been completed by the Late Paleozoic–Early Mesozoic. The Late Paleozoic glaciations may have created enough deep ocean circulation of cold, oxygen-laden water to result in complete ventilation, whereas earlier glaciations such as those of the Late Precambrian and Valanginian (Harland, 1974) and the Late Ordovician (Beuf *et al.*, 1971) appear to have led to reduction in anoxicity but not complete ventilation.

Development of relatively long-term, warm, equable conditions over much of the world at intervals in the Jurassic (Hallam, 1975, 1981) and Cretaceous appear to have resulted in oceanic anoxic events (Schlanger & Jenkyns, 1976; Arthur, 1979; Arthur & Schlanger, 1979; Thiede & van Andel, 1977; Ryan & Cita, 1977; Jenkyns, 1980). At this time, anoxic waters developed in the mid-ocean and spread across portions of the continental shelves. Schlanger and Jenkyns (1976, p. 182) suggested that oxygen deficient waters not only reached shelf sea depths during intervals in the Cretaceous, but that they also extended into deeper water.

Arthur (1979) reviewed Cretaceous and Tertiary paleoceanographic events and the sedimentary record. The Cenozoic stratigraphic record indicates that the Monterey Shale and related organic carbon-rich rocks in California as well as similar deposits in Chile and Japan (Garrison, 1975) reflect a widespread development of oxygen-poor waters over certain areas of continental slope and ocean basin during limited intervals of time in the Tertiary.

3. The Model

In their model of progressive ventilation of the ocean, Berry and Wilde (1978) drew attention to two primary ocean ventilating processes: (1) wind-induced mixing of the surface layer of the ocean, and (2) sinking of cold, oxygen-rich waters that formed at high latitudes accompanying sea-ice formation during glacial intervals. In the present-day, sea-ice is present permanently at the poles. One consequence is that the major factor in the formation of water masses is the wide variation in temperature of the world's oceans. Variation in open ocean salinity is relatively slight. Accordingly, most of the open ocean is cold and well oxygenated, reflecting the high latitude origin of deep and bottom waters. Salinity induced sinking of major parcels of ocean water occurs today only at mid-latitudes where evaporation exceeds precipitation. Mediterranean water in the Atlantic and Red Sea water in the Indian Ocean are typical of the consequences of high rates of evaporation in these marginal seas. A larger area of formation of higher salinity water is found polarward of the open ocean salinity maximum (Sverdrup *et al.*, 1942, p. 740). This water is typified by Central Water masses such as the Atlantic 18°C water (Worthington, 1959) which sinks at about 40°N and moves equatorward forming the subsurface salinity maximum. In the modern ocean, the volumes of water involved in such parcels are relatively small and not particularly significant when compared with the volumes of water involved in parcels that sink by temperature induced density.

In non-glacial times, however, when climates were relatively moderate worldwide and the temperature gradient from the poles to the equator was markedly less than at present, density differences reflecting salinity differences could have resulted in significantly large water parcels that sank to mid- and even deep ocean depths. As it is difficult if not impossible, to ascertain the actual temperature and evaporation–precipitation ratios during non-glacial periods, it seems appropriate to model such changes as a continuum and to define the conditions governing which combinations of temperature and salinity resulted in major water masses. Water masses are developed in the model herein that draw upon the interplay of suggested temperature and salinity regimes. The ventilation potential of these water masses is used to suggest oceanographic conditions during the Mesozoic and Cenozoic that could have resulted in a return of anoxic waters in the deep oceans, assuming that the oceans had been ventilated by the Late Paleozoic or Earliest Mesozoic. Under appropriate conditions of productivity and rising sea-level, certain of these anoxic waters could have spread onto continental shelves, as Schlanger and Jenkyns (1976) and Jenkyns (1980) noted had occurred in the Cretaceous, and Hallam (1981) indicated could have taken place in the mid-Jurrassic.

A means of illustrating the interrelationship of temperature and salinity on density is the Helland–Hansen (1916) or T–S diagram in which temperature and salinity are plotted linearly with density contoured on the plot. Density is given as Sigma-T:

$$\text{Sigma-T} = (\rho - 1) \times 1000 \tag{1}$$

where

$$\rho = \text{Density in gm}^{-3}$$

and

$$\text{Sigma-}t = f(\text{Temperature and Salinity})$$

(Cox, McCartney & Culkin, 1970).

Defant (1961, pp. 202–218) and Mamayev (1975) give detailed discussions of the use of T–S diagrams in the study of water masses and mixings. The intent here is to use T–S diagrams and modern T–S relationships to provide a framework to extrapolate and test the effects of non-glacial climates on the world ocean. Figure 1 shows such a diagram for hypothetical limits of water mass formation for two climates in non-glacial times: Case A; maximum non-glacial conditions with minimum thermal contrast between equatorial and polar regions; Case B; transition conditions either leading to or coming from glacial conditions, but without significant sea-ice formation.

The assumptions for these diagrams are as follows.

(1) The range of salinity variations and the mean salinity of the ocean has been essentially constant since early in the Paleozoic, as the fossil record indicates that certain marine stenohaline organisms such as crinoids, brittle stars, and starfish have survived with little change morphologically, and, apparently, in modes of life since the Early Palezoic.

(2) Wust (1936) calculated an empirical relationship between evaporation and precipitation which results in surface salinities. That relationship obtains since the Precambrian or:

$$\text{Surface salinity } (\%_{00}) = 34.60 + 0.0175 \ (\text{E–P}) \tag{2}$$

where

$$\text{E–P} = \text{Evaporation–precipitation in centimetres.}$$

This gives a mean salinity of 34.6‰ sufficiently close to the modern mean of 34.72‰ (Montgomery, 1958, p. 146).

(3) The maximum precipitation belts will always be equatorial and at high latitudes so that the minimum salinity by eqn (2) will be 33.55‰ using an Equatorial E–P = −60 cm (Wust, 1936). This defines a left-hand boundary to Figures 1 and 3. This boundary may shift right to higher salinities with decreasing temperatures as cooler air has less moisture capacity than warmer air.

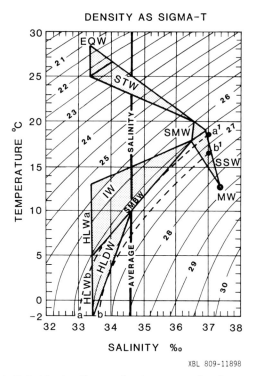

DENSITY AS SIGMA-T

SALINITY ‰

XBL 809-11898

Figure 1. Temperature/salinity/density diagram showing proposed major water masses in non-glacial times. Water mass abbreviations: EQW = Equatorial water, SSW = Shelf sea water, STW = Subtropical water, SMW = Salinity maximum water, HLW$_a$ = High latitude water for Case A conditions, HLW$_b$ = High latitude water for Case B conditions, IW = Intermediate water, SMBW = Salinity maximum bottom water, MW = Mediterranean water, a–a' gives the maximum density for Case A conditions, b–b' gives the maximum density for Case B conditions. For SSW water to be bottom water its volume must be sufficient to exceed the mixing ratio between SMW water and the appropriate SSW water given by the intersection with a–a' or b–b' (see Defant, 1961, p. 202 or Sverdrup *et al.*, 1942, pp. 143–146).

(4) Non-glacial times are defined by no or neglible sea-ice formation. The minimum temperatures would be above the freezing point of sea water or:

$$FP\ (^{\circ}C) = 0.036 - 0.0499\ \text{SAL}\ (\text{‰}) - 0.0001125\ \text{SAL}^2 \qquad (3)$$

(Fujino, Lewis & Perkins, 1974, p. 1797.) For 33.75‰ the freezing point is about −1.8°C.

(5) Due to the bi-modal distribution of maximum surface salinity between the minimum salinities at the equator and high latitudes, the maximum temperature at a given salinity would lie on a line (Figure 1) from maximum precipitation conditions at the equator (Equatorial Water (EQW)), to the surface salinity maximum area where salinity maximum water (SMW) is formed. This line would represent changes in latitude from 0° to about 40° in the Modern Ocean, or to the right of EQW (Figure 1). Polarward of salinity maximum water mass formation (to the left of SMW) precipitation increases. For the model, the high latitude maximum precipitation region begins at 13°C which is the warmest high latitude average temperature for the modern zone 40°–50° latitude (Sverdrup *et al.*, 1942, p. 127). Based on modern formation of high salinity 18°C water (Worthington, 1959), high latitude water (HLW) could form at any temperature below 18°C. The real value during non-glacial times would depend on the latitudinal (E–P) gradient.

(6) For volume considerations, the mixing of high salinity and low salinity waters are considered to produce ocean water of mean salinity to comply with assumption (2).

(7) The hypsographic curve (Kossinna, 1921; Menard & Smith, 1966) showing the distribution of depths in the ocean is effectively unchanged at the world-wide scale, since the beginning of the Paleozoic.

3.1. Case A

3.1.1. Water masses. Figure 1 gives the model T–S diagram for non-glacial conditions. To separate the environmental conditions of Case A from Case B, which is transitional to and from glacial conditions, we will define normal Case A times as having a thermal contrast and evaporation–precipitation gradient so that salinity maximum water (SMW) is a denser than high latitude water (HLW). This does not preclude denser shelf sea water but speaks only to the major oceanic water masses. Figure 1 shows that using 18°C water as a model salinity maximum water (Worthington, 1959), all HLW must be formed at 5°C or warmer.

A longitudinal section (Figure 2) shows the position of the water masses formed during these conditions. EQW would be the lightest water spreading polarward from the equator. SMW would be formed at the salinity maximum spreading both polarward and equatorward. SMW and EQW would spread equatorward and mix to form subtropical water (STW) which would sink under EQW. SMW spreading polarward would cool and reach a maximum density at 18°C (by analogy with 18°C water of Worthington (1959)). Some of the water would sink and maintain its identity as salinity maximum water (SMW). On the polarward margins the SMW would mix with HLW to form an intermediate water (IW) which would sink to a depth above pure SMW. Under special circumstances the coldest HLW can mix with SMW to produce bottom water at high latitudes (SMBW). Shelf sea water (SSW) forming at mid-latitudes in the zone of maximum evaporation would mix and accelerate the sinking of SMDW. However, due to the much larger volumes of open ocean water masses, the SSW would probably lose its identity as does Mediterranean and Red Sea Water in the Modern Ocean. By analogy, Figure 1 shows that mixing of 18°C (SMW) with Mediterranean Water (MW), a typical example of SSW, would produce denser water than SMBW for a ratio of SMW:MW less than 9:1.

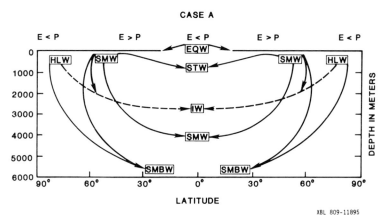

Figure 2. Longitudinal profile for Case A conditions. See Figure 1 for water mass designations.

3.1.2. Oxygen conditions. Figure 3 shows the Model T–S Pattern plotted with oxygen content contoured instead of density. Due to the strong inverse relationship between oxygen solubility and temperature (Weiss, 1977) the high latitude waters have appreciably higher initial oxygen content than low to mid-latitude waters. For modern conditions, oxygen consumption for 18°C water, which sinks only to 400 m, is 0.5 ml l^{-1} yr^{-1} (Worthington, 1959, p. 30). This indicates for an initial oxygen content of 5.35 ml l^{-1}, recycling to the surface or contact with another source of oxygen must be within 10 years. 18°C water presently occurs in the upper ocean where oxygen consumption is highest due to surface-produced organic oxygen

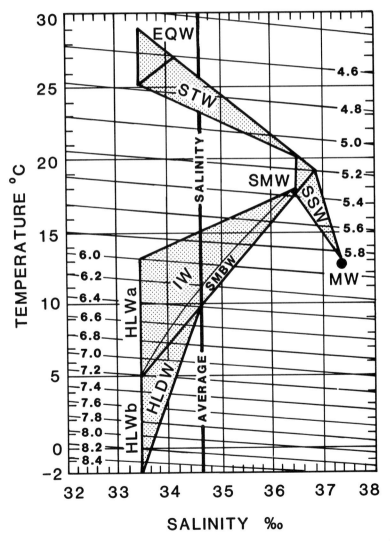

Figure 3. Temperature/salinity/oxygen saturation diagram with proposed major water masses in non-glacial times. See Figure 1 for water mass designations.

demand. However Munk (1966, p. 716) estimates that oxygen consumption in the deep ocean ranges from 0.0027 to 0.0053 ml l^{-1} yr^{-1}. For the mild climate of Case A, the minimum cycling times are given in Table 1. The actual rate of consumption with depth is described by Wyrtki (1962) as

$$R = R_0 \exp{(-\alpha Z)} \qquad (4)$$

where:

  R = rate of consumption at depth Z

  R_0 = surface consumption which is a function of local productivity

  α = attenuation coefficient related to properties of oxidizable substances available

  Z = depth.

However for this model the mean values as noted will be used in lieu of the precise values needed to solve eqn (4).

3.2.   Case B

3.2.1.   *Water masses.* As noted above, ᴗase B is indicative of transitional conditions entering or leaving glacial times. That is, high thermal contrast, but without significant sea-ice formation. Figure 1 gives the proposed T–S relationships and Figure 4 shows the longitudinal distribution of the water masses. The major difference is the addition of water mass formation below 5°C with bottom and deep water formed at high latitudes rather than from salinity maximum water at the mid-latitude. The modern analogy for this low salinity and temperature water is Antarctic Intermediate water (Sverdrup *et al.*, 1942, p. 619). The latitude of temperatures from 13°C to 5°C shift equatorward so that waters called HLW$_a$ in Case A are on a continuum to colder high latitude waters. HLW$_b$ is reserved for low salinity waters from 5°C to −2°C. The sinking of HLW$_b$ to form deep and bottom waters also converts the salinity maximum bottom water (SMBW) of Case A to a new Intermediate Water. To avoid confusion the old designation SMBW is maintained although it is no longer the densest water mass. Conditions above 5°C would be the same as Case A except for an equatorward shift. However, it seems likely that the

Table 1. Oxygen conditions

Water mass	Maximum initial O$_2$ Content (ml l^{-1})	Cycling time before depletion (years)[1]	(years)[2]
Equatorial water (EQW)	4.5–4.8	1125–1200	9
Salinity maximum water (SMW)	5.1–5.35	1275–1338	10
Intermediate water (IW) at average salinity	5.8–6.3	1450–1575	10.2–12.6
Salinity maximum bottom water (SMBW) at average salinity	6.3–6.5	1575–1625	
High latitude water (HLW)			
at 13°C	6.0	1500	
at 5°C	7.2	1800	
at −2°C	8.4	2100	
High latitude deep water	6.5–8.4	1625–2100	

[1] Oxygen consumption (deep ocean): Mean value of 0.004 ml l^{-1} yr^{-1}
                                    Range 0.0027 to 0.0053 ml l^{-1} yr^{-1}
                                    (Munk 1966, pp. 716)
[2] Oxygen consumption (surface): 0.5 ml l^{-1} yr^{-1} (Worthington, 1959, p. 302)

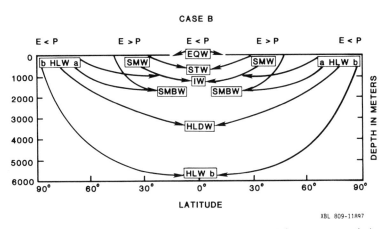

Figure 4. Longitudinal profile for Case B conditions. See Figure 1 for water mass designations.

major water mass formation would shift to the lower temperature regions of HLW_b as the greater density would indicate a higher rate of formation. With extended cold temperature regions of water formation SMW cannot mix directly with the new HLW_b as SMBW intervenes. Accordingly, HLDW would be a mixture of SMBW and HLW_b. The mixing line between 18°C SMW and 5° HLW would be analogous to the Central Water of the Modern Ocean. The maximum open ocean density increases to 27.2 Sigma-T so that the mixing ratio for SMW/SSW as typified by MW is now about 3:2 making it even less likely that SSW could penetrate to deep depths before losing its characteristics by mixing.

3.2.2. Oxygen conditions. Figure 3 shows the additional oxygen capacity of HLW_b compared to Case A conditions. With the confinement of Case A waters to shallower depths by the sinking of colder high latitude water, the chances for Case A waters cycling back to the surface or to be in the wind mixed layer are enhanced.

3.3. Volumes

The amount of water of given characteristics for the specified climatic conditions is a function of the rate of formation, the area of formation, and the residence time of the water mass in the ocean. A maximum value for modern rates of formation is given by Munk (1966) caused by the freezing of the Antarctic ice pack as 9×10^{20} g yr^{-1} or 9×10^{17} l yr^{-1}. At this rate the entire volume of the ocean 1.369×10^{21} l (Montgomery, 1958) could be formed in about 1.5×10^3 years. However, it is unlikely that sufficient data can be derived from the geologic record to calculate volumes in this direct manner.

Another method to estimate volumes is to use the assumption of constant average salinity and to use the T–S plot to determine what mixing ratios (volumes) between water masses will maintain the average salinity of the ocean. Assume for non-glacial times the two major water masses are SMW (High Salinity) and HLW (Low Salinity), ignoring volumetrically EQW (Low Salinity) and SSW (High Salinity). Thus, the mixing ratio of SMW/HLW is 0.36 or 36% SMW formed to 64% HLW (Figure 1).

The hypsographic curve (Kossina, 1921, and Menard & Smith, 1966) can be used to estimate the range of depths occupied by the water masses for Case A and B climates. The hypsographic curve is given as a function of area and depth so

$$V = \int dA dZ \tag{5}$$

where

$$V = \text{Volume}$$
$$A = \text{Surface Area}$$
$$Z = \text{Depth}$$

Integration of the hypsographic curve to give the oceanic volumes as a function of depth is given in Appendix A. Figure 5 shows the hypsographic curve for volume as well as for the conventional oceanic area, with the volumes of each major water mass plotted based on the T–S mixing ratios. The intersection of the volume percentage and the volume curve gives the depth to which that water mass as bottom water would fill the ocean basin. Replotting the depth ranges on the area hypometric curve (Figure 5) gives the bottom area intersected by the water masses. By analogy with modern conditions using a wind-mixed surface layer of 200 m, the base of the pycnocline would be 2900 m for Case A and 1500 m for Case B. These would be minimum depths as consideration of other water mass production would shift the base to deeper depths; although the ratio of SMW:HLW would still be 36:64.

3.4. Geologic implications of the model

The circulation models for Case A and B can be used to estimate the oxygen conditions with depth and latitude with the addition of information on the magnitude and distribution of organic matter in the ocean. As a first approximation, high productivity is found in zonal belts at the Equator and today at about 60° N and S at the areas of planetary upwellings. Minimum productivity is found at mid-latitudes at the planetary oceanic convergences at the approximate latitude of the salinity maxima.

For the climatic conditions of Case A, with the general lower oxygen saturation values, the ventilation of the ocean is the poorest. Figure 6 shows the Case A

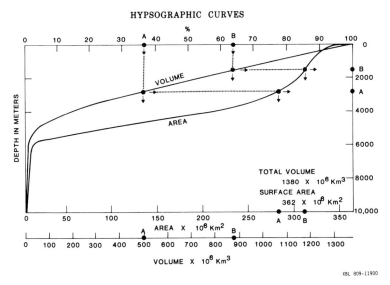

XBL 809-11900

Figure 5. Hypsographic curves for volume and area of the world ocean-area data from Kossinna (1921). See Appendix A for volume calculations.

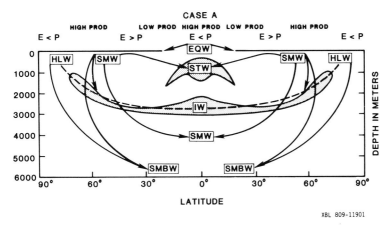

Figure 6. Longitude profile for Case A conditions with oxygen depleted zones shaded. See Figure 1 for water mass designations.

longitudinal section with potential oxygen depleted depths shaded. The high production at the Equator would produce a shallow depleted zone below the mixed layer 200 m to a depth at which the surface organic matter would be oxidized. A deeper depleted region would be found along the base of the Case A pycnocline with its origin in the productivity maxima at high latitudes. Because of the deep pycnocline of Case A, the residence time of the water sinking along the pycnoclinal density is long so that water continues to be depleted in oxygen. There would be a mid-pycnoclinal aerated zone below the equatorial depleted zone and above the depleted layer at the base of the pycnocline as this water has its origin in the low productivity areas of mid-latitude and would have a low oxygen demand. Whether the depleted zones would be anoxic is a function of the residence time of the water masses and contact with other oxygenated waters. For Case A, with its initial low oxygen saturation values compared to the Modern Ocean, it seems likely that the oxygen-depleted areas indeed would be anoxic. Geologically, it would mean at the Equator a return to anoxic conditions below 200 m seen in the Paleozoic and that the outer continental shelf and upper slope would be anoxic. A deep Equatorial anoxic zone explains the anoxic events of Schlanger and Jenkyns (1976) and Thiede and van Andel (1977) in the Cretaceous. As seen in Figure 3, the maximum depletion would occur at the mildest climates and the lowest temperature contrast between the salinity maximum and high latitudes.

For Case B (Figure 7), the oxygen-depleted zone would follow the shallower base of the pycnocline. The two Equatorial depleted zones of Case A (Figure 6) would merge for Case B. With the strong dependence between oxygen solubility and temperature, the lower temperatures of Case B indicate a lesser probability of actual anoxic conditions except in the Equatorial Belt.

3.4.1. Stratification and potential overturn. In both Cases A and B the oceans are density stratified except for the conditions for 5°C for high latitude water mass formation, where the density of HLW and SMW are the same. The stability factor (Defant, 1961, pp. 195–201) is a function not only of the density distribution with depth, but also the compressibility and the adiabatic effect so that simple low density contrast in a deep ocean is not sufficient to initiate Benard type convection or overturn. However, the oceanic climatic boundary conditions between Case A and B certainly suggest potential neutral stability conditions. The possibility for periodic overturn of oxygen-depleted mid-waters into the upper pycnocline and the mixed

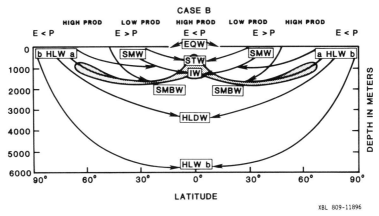

Figure 7. Longitudinal profile for Case B conditions with oxygen depleted zones shaded. See Figure 1 for water mass designations.

layer has interesting geologic and evolutionary implications. Such conditions occur with a reasonably large temperature contrast between 18°C for SMW and 5°C for HLW, rather than for small temperature contrast, as one intuitively would expect. This paradox is due to the conservation of the average salinity and the change in surface salinity related to the zonal change in evaporation minus precipitation.

The potential for overturn conditions thus would occur not only as a precursor of the onset or cessation of true glacial times (freezing of sea-ice); but also during non-glacial times which are cooler than Case A conditions.

4.   Conclusions

A model of possible Mesozoic oceans water masses is developed herein that follows from the model of progressive ocean ventilation through the Paleozoic (Berry & Wilde, 1978). If the oceans were ventilated by the Late Paleozoic–Earliest Mesozoic, as the stratigraphic record seems to suggest, then the development of certain widespread, coeval black shales in the ocean basin stratigraphic record for intervals in the Jurassic (Hallam, 1981) and Cretaceous (the deposits of the Cretaceous Oceanic Anoxic Events of Schlanger and Jenkyns (1976), Arthur and Schlanger (1979), and Jenkyns (1980)) suggests a return of anoxic mid-ocean waters during short time intervals in the Mesozoic. Potentially, the Cenozoic Monterey Shale and equivalent units is indicative of relatively widespread but short interval of anoxic waters in the Cenozoic.

The water masses described herein are based upon examples from modern oceans and, therefore, developed from realistic temperature, salinity, and density values. Under climatic conditions in which mild, equable climates were widespread (Case A), bottom waters could have formed at the surface in mid-latitudes, near salinity maxima. Under climatic conditions of cold polar area (but lacking in sea-ice) (Case B), bottom waters would have been formed at high latitudes.

For Case A climatic conditions, those in which high latitude temperatures were above 5°C, two oxygen-depleted layers formed in equatorial regions. One extended from about 200 m to the depth of completely oxidized organic material that originated at the surface. The second extended along the base of the pycnocline at about 2900 m. It rose in high latitudes as a response to high seasonal productivity in those areas. As climates cooled in Case B, the two low oxygen water parcels diminished. The lower parcel in essence rose to join with a much smaller upper parcel (see Figures 4 and 7). During Case B conditions or cooler, the former anoxic waters of Case A probably became oxygen minimum water. Return of equable climates and increased productivity locally could bring back anoxic conditions. If climates were equable for a long time, then Case A climates would return and anoxic waters could be re-established. Climatic fluctuations are suggested to have resulted in the sequences of oceanic anoxic events of the Jurassic and Cretaceous (and those of the Cenozoic, if such existed).

The presence of an upper parcel of anoxic water during mild, equable climatic conditions (Case A) (Figure 6), could have led to the spread of anoxic waters over continental shelves of the time, provided that sea-levels rose. If they did not, then the mid-ocean anoxic waters would not have reached the shelves.

If only density of the water masses is considered, transition from Case A to Case B conditions could result in an essentially neutral stratification. If that condition did occur, then the potential for oceanic overturn seems to have existed. This potential would have developed at times when the density difference between major water masses formed at high latitude and those formed at mid-latitudes was slight or negligible. Overturn could have had an enriching effect on the surface waters and been followed, therefore, by increased productivity. Alternatively, if a large volume of anoxic water was brought near the surface, overturn could have resulted in a major extinction. Potentially, such a mass killing could have been followed by heightened productivity as waters freshened and nutrients were plentiful.

The postulated water masses are based upon modern water masses and their formation, as well as from extrapolation of conditions based on the Mesozoic stratigraphic and climatic record. If intervals of equable climates did persist in the Mesozoic and if these intervals were preceded and succeeded by those of colder conditions, then the oceanic water masses discussed herein may provide an explanation for periodic formation of widespread, time-synchronous organic carbon-rich rocks in the Mesozoic (and, possibly the Cenozoic).

In any event, the properties of any hypothetical water mass derived from paleo-oceanographic of climatologic arguments must conform to the basic constraints of Temperature–Salinity diagrams as amply discussed by Mamayev (1975). Accordingly, the authors hope the above model and analysis will alert the geologic community to the advantages of Temperature–Salinity diagram as a useful tool to investigate and illustrate ancient marine conditions.

Appendix A: Volumetric hypsographic curve calculations

For

$$V = \int dA\, dZ$$

Δ volume $= \Delta$ depth $\times$ average area

or

$$V_i - V_{i-1} = \left(Z_i - Z_{i-1}\right)\left(\frac{A_i + A_{i-1}}{2}\right)$$

$$= \Delta Z\left(A_i - 1 + \frac{\Delta A_i}{2}\right)$$

let

$$\Delta Z_i = \text{interval from } A_i \text{ to } Z_{i-1}$$

$$\Delta Z = Z - Z_{i-1}$$

then

$$\Delta V = \Delta Z \left(A_{i-1} + \frac{\Delta A_i}{2} \times \frac{\Delta Z}{\Delta Z_i} \right)$$

Table A1. Computations for Figure Hypsographic Curves. Data for Areas from Kossinna (1921).

Depth (m)	Area[1]	%	Comm. Area*	Comm. (%) Area	Volume*	%	Comm. Volume*	Comm. (%) Volume
10 000	0	0	0	0	0	0	0	0
6 000	4.5	1.2	4.5	1.2	9	0.6	9	0.6
5 000	84	23.2	88.5	24.4	46.5	3.3	55.5	3.9
4 000	120	33.1	208.5	57.5	148.5	10.8	204.0	14.7
3 000	70	19.3	278.5	76.8	243.5	17.6	447.5	32.3
2 000	24.5	6.8	303	83.6	290.75	21.1	738.25	53.4
1 000	15	4.1	318	87.6	301.5	212.5	1049.75	75.9
200	15.5	.3	333.5	92.0	260.6	18.9	1310.35	94.8
0	28.5	7.9	362	99.9	69.55	5.0	1379.9	99.8

* Units: Area—10^6 km²; Volume—10^6 km³

References

Arthur, M. A. 1979. Paleooceanographic events—recognition, resolution, and reconsideration. *U.S. National Report*, 17th Assembly, International Union of Geological Sciences 17 (7), 1474–1494.

Arthur, M. A. & Natland, J. H. 1979. Carbonaceous sediments in the North Sea and South Atlantic: The role of salinity in stable stratification of Early Cretaceous Basins. (Eds. M. Talwani, W. Hay and W. B. F. Ryan) *Deep Drilling Results in the Atlantic Ocean, Continental Margins and Paleoenvironments. Maurice Ewing Series* 3, American Geophysical Union, pp. 375–401.

Arthur, M. A. & Schlanger, S. O. 1979. Middle Cretaceous "oceanic anoxic events" as causal factors in development of reef reservoired giant oil fields. *American Association of Petroleum Geologists Bulletin* 63, 870–885.

Bernouilli, D. & Jenkyns, H. C. 1974. Alpine, Mediterranean, and Central Atlantic Mesozoic facies in relation to the early evolution of the Tethys. Modern and Ancient Geosynclinal Sedimentation. (Eds. R. H. Dott & R. H. Shaver) *Society of Economic Petroleum and Mining Special Publication* 19, pp. 129–160.

Berry, W. B. N. 1974. Facies distribution patterns of some marine benthic faunas in Early Paleozoic platform environments. *Palaeogeography, Palaeoclimatology, Palaeoecology* 15, 153–168.

Berry, W. B. N. & Boucot, A. J. 1968. Continental development from Silurian viewpoint. *International Geological Congress 23rd, Prague 1968, Proceedings Section 3, Orogenic Belts*, pp. 15–23.

Berry, W. B. N. & Boucot, A. J. 1973. Glacio-eustatic control of Late Ordovician—Early Silurian platform sedimentation and faunal changes. *Geological Society of America Bulletin* 84, 275–284.

Berry, W. B. N. & Wilde, P. 1978. Progressive ventilation of the oceans—an explanation for the distribution of the Lower Paleozoic black shales. *American Journal Science* 278, 257–275.

Beuf, S., Biju-Duval, B., de Charpal, O., Rognon, P., Gariel, O. & Bennacef, A. 1971. Les gres du Paleozoique inferieur au Sahara—sedimentation et discontinuities, evolution structurale d'un craton. *Institut Francais du Petroleum*, Technipub, Paris. 464 pp.

Borchert, H. 1965. Formation of marine sedimentary iron ores. *Chemical Oceanography* (Eds. Riley & Skirrow), Vol. 2. London: Academic Press, pp. 159–204.

Chamberlain, T. C. 1906. On a possible reversal of deep-sea circulation and its influence on geologic climates. *Journal of Geology* 14, 371–372.

Cloud, P. 1976. Beginnings of biospheric evolution and their biogeochemical consequences. *Paleobiology*, 2, 351–387.

Cook, H. E. & Taylor, M. E. 1977. Comparison of continental slope and shelf environments in the Upper Cambrian and lowest Orovician of Nevada. Deep water carbonate environments, (Eds. H. E. Cook and P. Enos) *Society of Economic Paleontologists and Mineralogists Special Publications* 25, pp. 51–81.

Cowie, J. W. 1974. The Cambrian of Spitsbergen and Scotland. *Cambrian of the British Isles, Norden, and Spitsbergen* (Ed. C. H. Holland) Chichester: Wiley, pp. 123–155.

Cox, R. A., McCartney, M. H. & Culkin, F. 1970. The specific gravity/salinity/temperature relationship in natural sea water. *Deep-Sea Research* **17**, 679–689.

Dean, W. 1980. The Ordovician System in the Near and Middle East: Correlation Chart and explanatory notes. *International Union of Geological Sciences Publication* **2**, 22 pp.

Defant, A. 1961. *Physical Oceanography*. Oxford, Pergamon, 729 pp.

Fujino, K., Lewis, E. L. & Perkin, R. G. 1974. The freezing point of sea water at pressures up to 100 Bars. *Journal of Geophysical Research* **79**, 1792–1797.

Garrison, R. E. 1975. Neogene diatomaceous sedimentation in East Asia: a review with recommendations for further study. *Technical Bulletin, Co-ordinating Committee Economic and Social Community (Asia and the Pacific) (United Nations)* **9**, 57–69.

Glaeser, J. D. 1979. Catskill delta slope sediments in the central Appalachian Basin: Source deposits and reservoir deposits. *Geology of Continental Slopes, Society of Economic Paleontologists and Mineralogists Special Publication* **27**, 343–357.

Hallam, A. 1975. *Jurassic Environments*. Cambridge: Cambridge University Press, 269 pp.

Hallam, A. 1981. *Facies Interpretation and the Stratigraphic Record*. San Francisco: W. H. Freeman, 291 pp.

Harland, W. B. 1974. The Pre-Cambrian–Cambrian boundary. *Cambrian of the British Isles, Norden, and Spitsbergen*, (Ed. C. A. Holland), Chichester: Wiley, pp. 15–42.

Heckel, P. H. & Witzke, B. J. 1979. Devonian world palaeogeography determined from distribution of carbonates and related lithic palaeo-climatic indicators. *The Devonian System* (Eds. M. R. House, C. T. Scrutton & M. G. Bassett) *Special Papers in Palaeontology* **23**, 99–123.

Helland-Hansen, B. 1916. Nogen hydrografiske metoder. *Skandinaviske Naturforskeres Moede Forhandlinger*, 357–359.

Holland, C. H. (Ed.) 1971. *Cambrian of the New World*. New York: Wiley Interscience, 456 pp.

Jenkyns, H. C. 1978. Pelagic environments. In H. G. Reading, Ed., *Sedimentary Environments and Facies*. Oxford: Blackwell Scientific Publications, 314–371.

Jenkyns, H. C. 1980. Cretaceous anoxic events: from continents to oceans. *Journal of the Geological Society of London* **137**, 177–188.

Kossinna, E. 1921. Die Tiefen des weltmeeres. Institute Meereskunde, Veroffentlichung, *Geographische–Naturwissenschaft* **9**, 1–70.

Krebs, W. 1979. Devonian Basinal Facies. *The Devonian System. Special Papers in Palaeontology* **23**, 123–139.

Leggett, J. K. 1980. British Lower Palaeozoic black shales and their palaeo-oceanographic significance. *Journal of the Geological Society of London* **137**, 139–156.

Mamayev, O. I. 1975. *Temperature–salinity Analysis of World Ocean Waters*. Amsterdam: Elsevier, 374 pp.

Martinsson, A. 1974. *The Cambrian of Norden. Cambrian of the British Isles, Norden, and Spitsbergen* (Ed. C. H. Holland). Chichester: Wiley, pp. 185–283.

Menard, H. W. and Smith, S. M. 1966. Hysometry of ocean basin provinces. *Journal of Geophysical Research* **71**, 4305–4325.

Montgomery, R. B. 1958. Water Characteristics of Atlantic Ocean and of World Ocean. *Deep-Sea Research* **5**, 134–148.

Munk, W. 1966. Abyssal recipes. *Deep-Sea Research* **13**, 707–730.

Ross, C. A. 1979. Evolution of fusulinacea (Protozoa) in Late Paleozoic space and time. *Historical Biogeography, Plate Tectonics, and the Changing Environment*, (Eds. J. Gray and A. J. Boucot), Corvallis, Oregon: Oregon State University Press, pp. 215–226.

Rushton, A. W. A. 1974. The Cambrian of England and Wales. In *Cambrian of the British Isles, Norden, and Spitsbergen*, (Ed. C. H. Holland), Chichester: Wiley, London, 43–121.

Ryan, W. B. F. & Cita, M. B. 1977. Ignorance concerning of ocean-wide stagnation. *Marine Geology* **23**, 197–215.

Schlanger, S. O. & Jenkyns, H. C. 1976. Cretaceous oceanic anoxic events: causes and consequences. *Geologie en Mijnbouw* **55**, 179–184.

Stewart, J. H. & Poole, F. G. 1974. Lower Paleozoic and Uppermost Precambrian Cordilleran miogeocline, Great Basin, western United States. In W. R. Dickinson, Ed., *Tectonics and Sedimentation, Society of Economic Paleontologists and Mineralogists Special Publication* **22**, 28–57.

Stewart, J. H., MacMillan, J. R., Nichols, K. M. & Stevens, C. H. 1977. Deep-water upper paleozoic rocks in north-central Nevada—a study of the type area of the Havallah formation. *Paleozoic Paleogeography of the Western United States: Pacific Coast Paleogeography Symposium I. Pacific Section, Society of Economic Paleontologists and Mineralogists*, 337–347.

Sverdrup, H. V., Johnson, M. W. & Fleming, R. H. 1942. *The Oceans*. Englewood Cliffs, New Jersey: Prentice-Hall, 1042 pp.

Thiede, J. & van Andel, T. H. 1977. The paleoenvironment of anaerobic sediments in the Late Mesozoic South Atlantic Ocean. *Earth and Planetary Science Letters* **33**, 301–309.

Weiss, R. F. 1970. The solubility of nitrogen, oxygen, and argon in water and sea water. *Deep-Sea Research* **17**, 721–725.

Worthington, L. V. 1959. The 18°C water in the Sargasso Sea. *Deep-Sea Research* **5**, 297–305.

Wust, G. 1936. Oberflachensalzgehalt, Verdunstung, und Neiderschlag auf dem Weltmeer. *Festscrift Norbert Krebs. Landerkundliche Forschung, Stuttgart*, 347–359.

Wyrtki, K. 1962. The oxygen minimum in relation to ocean circulation. *Deep-Sea Research* **9**, 11–23.

Subject Index